The Open University

S267 HOW THE EARTH WORKS: THE EARTH'S INTERIOR

BLOCK 2

HOW PLATE TECTONICS WORKS

Prepared for the Course Team by
ANDREW BELL

THE S267 COURSE TEAM

CHAIRMAN

Peter J Smith

C J Hawkesworth

COURSE COORDINATOR

Veronica M E Barnes

COURSE MANAGER

Val Russell

AUTHORS

Andrew Bell

Stephen Blake

Nigel Harris

David A Rothery

Hazel Rymer

EDITORS

David Tillotson

Gerry Bearman

Sue Glover

DESIGNER

Caroline Husher

GRAPHIC ARTIST

Alison George

BBC

David Jackson

The Open University, Walton Hall, Milton Keynes, MK7 6AA.

First published 1993.

Edited, designed and typeset in the United Kingdom by the Open University.

Printed in the United Kingdom by Thanet Press Ltd. Margate. Kent.

ISBN 0 7492 8163 4

This text forms part of an Open University Second Level Course. If you would like a copy of *Studying with the Open University*, please write to the Central Enquiry Service, PO Box 200, The Open University, Walton Hall, Milton Keynes, MK7 6YZ. If you have not already enrolled on the Course and would like to buy this or other Open University material, please write to Open University Educational Enterprises Ltd, 12 Cofferidge Close, Stony Stratford, Milton Keynes, MK11 1BY, United Kingdom.

3.1

S267b2i3.1

S267
HOW THE EARTH WORKS:
The Earth's Interior

BLOCK 2
HOW PLATE TECTONICS WORKS

2.1	**INTRODUCTION**	**3**
2.2	**INVESTIGATING PLATE MOTION**	**7**
2.2.1	Relative plate movements	17
2.2.2	Towards true plate motions	19
2.2.3	Measuring modern plate movements	27
2.2.4	How old are oceanic plates?	28
	Summary of Section 2.2	29
	Objectives for Section 2.2	30
	SAQs for Section 2.2	31
2.3	**OCEAN PLATE BOUNDARIES**	**33**
2.3.1	Constructive plate boundaries	33
2.3.2	Transform faults and fracture zones	43
2.3.3	Destructive plate boundaries	46
2.3.4	Combined plate boundaries	57
2.3.5	Triple junctions	58
	Summary of Section 2.3	60
	Objectives for Section 2.3	61
	SAQs for Section 2.3	62
2.4	**CHANGING OCEAN PLATE BOUNDARIES WITH TIME**	**63**
2.4.1	Changes at spreading ridges with time	63
2.4.2	Plate boundary changes due to subduction	65
	Summary of Section 2.4	68
	Objectives for Section 2.4	68
	SAQs for Section 2.4	68
2.5	**PLATE TECTONICS AND CONTINENTAL CRUST**	**69**
2.5.1	How continental crust differs from oceanic crust	69
2.5.2	Constructive margins within continental crust	70
2.5.3	Destructive margins at the ocean–continent boundary	75
2.5.4	Destructive margins and continental collision	78
2.5.5	Conservative margins within continental crust	82
2.5.6	Extension work	83
	Summary of Section 2.5	84
	Objectives for Section 2.5	85
	SAQs for Section 2.5	86

2.6 **WHAT DRIVES THE PLATES?**   **87**

    2.6.1   Forces acting on the bottom of plates   87

    2.6.2   Forces acting at plate margins   88

    2.6.3   Adding plate forces together   90

    Summary of Section 2.6   94

    Objectives for Section 2.6   95

    SAQs for Section 2.6   95

ITQ ANSWERS AND COMMENTS   **97**

SAQ ANSWERS AND COMMENTS   **105**

SUGGESTIONS FOR FURTHER READING   **108**

ACKNOWLEDGEMENTS   **109**

BLOCK 2 COLOUR PLATE SECTION

2.1 INTRODUCTION

When you start reading this Block, you will probably have a basic grasp of plate tectonics. You will know that the Earth's outer shell is composed of plates of rock material which move relative to one another and over the Earth's surface. This is the fundamental point of the theory of plate tectonics, which has revolutionized Earth science in the latter part of the 20th century, and provides an overall tectonic framework that explains the movement patterns of the planet's outermost rock layer. It has also fundamentally changed the way we think about how the Earth works. It's a relatively recent development, having been stated and developed almost entirely since the mid-1960s. The aim of this Block is to show how the outermost shell works.

We have already referred to plates as if there were no possible doubt that they exist. It should be made clear from the outset that geoscientists *infer* that plates exist, based on deductive reasoning from an extremely large body of scientific information. That vast geoscientific database, which incidentally is growing at an alarming rate, provides ample evidence for plate tectonics. This Block tries to present some of that evidence in an ordered form. Once you have read it, we hope that you will be able to understand why the concept dominates modern Earth science.

Section 2.2 details how geoscientists can tell that plates move, both relative to each other and with respect to the Earth's axis of rotation. In an extended discussion, Section 2.3 describes the three fundamental types of oceanic plate boundary and their principal geophysical and tectonic features. Section 2.4 examines how plate tectonic boundaries within the oceans change with time, while Section 2.5 introduces the additional complexities of plate tectonic boundaries within continental crust, developing the theme that mountain belts are related to present or past plate activity. Lastly, Section 2.6 discusses the forces that drive the plates.

Throughout the Block, we refer to material in the Colour Plate Section (CPS) and to features and places seen on the map entitled *This Dynamic Planet — World Map of Volcanoes, Earthquakes and Plate Tectonics*, which we call the Smithsonian Map. You will need this to hand while you study the Block. The Smithsonian Map is introduced and analysed by two audio tapes (AV 05 'Smithsonian Map, Part I', and AV 06 'Smithsonian Map, Part II'. The first of these outlines the basic features of the map while the second puts these features into their plate tectonic context. You should listen to the first of these tapes at the beginning of the Block and the second at the end.

Block 2 also has one video programme associated with it, VB 03 'Plates in Motion: The San Andreas Fault', which you should view when asked to near the end of the Block. There are no specimens or experiments within the Home Kit that relate specifically to this Block.

In an overall context, Block 2 forms the global tectonic framework on which Blocks 3 and 4 are built, and contrasts with those of other worlds in Block 5. It occupies a central position in the Course. You should study it in the sequence in which it was written, and make sure you understand the basic principles of plate tectonics (using the SAQs) which it develops before moving on to other Blocks.

Block 2 contains, apart from this introduction, five main Sections with the following approximate Course-unit equivalents (CUEs):

Section	CUE
2.2	0.7
2.3	0.8
2.4	0.2
2.5	0.5
2.6	0.3
Total	2.5

At the start of our studies of the tectonics of plates, we need to review our existing knowledge. Listen first to AV 05 'Smithsonian Map, Part I', running time 36 minutes. (*Note*: There are no AV notes associated with AV 05.)

We can summarize the ideas (introduced both in Block 1 and in AV 05) about the most important ways by which plates are recognized:

- Plates are distinct regions of relatively rigid lithosphere, moving relative to one another.

- The extent of individual plates is established by recognizing their boundaries.

- We can also recognize plate boundaries by looking for signs of tectonic activity, as indicated by earthquakes and volcanoes.

- We can help to recognize plate boundaries by looking for regions of anomalous heat flow.

- Earthquakes are generated by movement of bodies of rock against each other across faults. The pattern of earthquake epicentres defined around the Pacific Ocean indicates tectonic activity and suggests that the Pacific margin is a plate boundary.

- Volcanically active continental margins, such as those that border the Pacific Ocean, are likely candidates for tectonic activity, and hence plate boundaries.

- The seismically active fault zones around the Pacific are also zones of increased heat flow, so high heat flow could be diagnostic of plate boundaries. However, many of the relatively inactive ridges and seamounts associated with 'hot spots' are also sites of increased heat flow. Therefore, high heat flow alone does not locate plate boundaries.

- Some continental margins, such as those that border the Atlantic Ocean, are tectonically quiet and are unlikely to represent plate boundaries.

- Bathymetrically high ridges within the oceans are sometimes tectonically active. Some of these are identified as divergent boundaries on the Smithsonian Map, and are obvious candidates for plate boundaries. However, other distinct ocean ridges are either associated primarily with volcanoes and seamounts, such as the Hawaiian Ridge, or are seismically and volcanically inactive, such as the Ninety-East Ridge. Not all ocean ridges are plate boundaries.

- Long linear mountain belts within or bordering the continents are clearly shown on the Smithsonian Map. These are often tectonically active and may therefore represent either present or former plate boundaries.

- Large areas of low-lying continents between mountain belts are also obvious on the Smithsonian Map and have been tectonically stable for long periods of geological time. They are unlikely to contain any currently active plate boundaries and so must represent the interior of modern-day plates. By analogy, the tectonically stable parts of the ocean floor probably also represent the interiors of plates.

- The theory of plate tectonics proposes (a) that the seismically and volcanically active linear belts locate the plate boundary network, and (b) that the cool, inactive interiors of both oceans and continents represent the interiors of the tectonic plates themselves.

- Oceanic plates are composed of up to 7–11 km of largely igneous rock in a layered form. These crustal rocks are separated from mantle rocks by a seismic discontinuity called the Moho.

- Continental plates are much thicker, more compositionally variable and more silica-rich than oceanic plates. The Moho can be recognized under continental crust at typical depths of 35–50 km, much deeper than under the oceans.

With these basic points fresh in our minds, let us turn now to the question of identifying and understanding plate tectonic processes.

2.2 INVESTIGATING PLATE MOTION

The Introduction that accompanies the Smithsonian Map tells us that '... the lithosphere is made up of a dozen or so rigid slabs that are moving horizontally at speeds from 1 to over 10 cm each year'. In other words, the plates are moving. How do we know this? Do all the plates move, or only some of them? Do they move at constant speeds or does this vary? Do they move in constant directions or does that vary, too?

Geometric continental reconstructions

We can show that plates move if we can identify the movement of the continents that form part of some of the plates. The simplest way of doing that is to examine the way they fit together geometrically. The geometric match between the coastlines of America and Africa was noted as early as 1620 by Sir Francis Bacon, although he believed that a central ocean had foundered rather than that the continents drifted apart. That continents might indeed have drifted apart was mentioned by Professor Theodore Lilienthal in a work dated 1756, but the concept of actively drifting continents was not developed until around 1910. The first computer-drawn reassembly of the continents around the Atlantic was performed by Sir Edward Bullard and colleagues in 1965 (Figure 2.1). Bullard fitted Africa with South America, North America and Western Europe based on plates that formed parts of the surface of a sphere which he fitted together at the 500 fathom (*c.* 1 000 m) contour. He included all the shallow-water shelves that surround the continental blocks as part of the continents themselves.

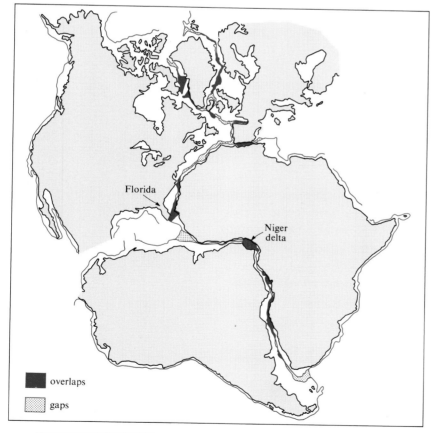

Florida

Niger delta

overlaps

gaps

Figure 2.1 Fit of the continents surrounding the Atlantic, generated by matching the continental shape at 500 fathoms.

ITQ 1

Figure 2.1 shows some gaps and overlaps in the continental fit around the Atlantic. If continents had once been joined, there should ideally be no gaps or overlaps. Why might overlaps exist?

The procedure performed by Bullard and his colleagues used concepts in spherical geometry that need not detain us here. We need only note that the fit at almost 1000m depth on Figure 2.1 is very good, with few gaps and overlaps. Most of the overlaps can be traced to features that developed after the continents split apart, such as the Bahama coral platform in southeastern North America and the recent sediments of the Niger delta off western Africa. Iceland is also absent from the North Atlantic, but we know from radiometric dates that Iceland developed as a volcanic island complex after the North Atlantic had started to open.

Bullard's technique proved so effective in generating a pre-drift picture of the Atlantic region that it stimulated further attempts to fit other continental masses together. In particular, the southern Atlantic continents were fitted with India, Australasia and Antarctica by A. G. Smith and A. Hallam in 1970, raising the possibility that almost all of the continental land-masses were once joined together to form a 'supercontinent' (Figure 2.2).

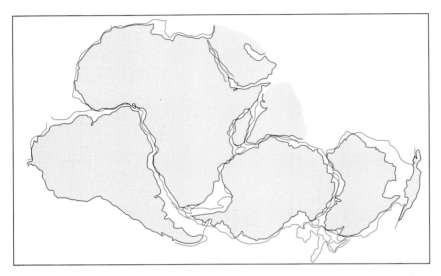

Figure 2.2   The best fit of the Southern Ocean continents of South America, Africa, India, Antarctica and Australasia.

Unfortunately, the fit of the Southern Ocean continents is not as clear-cut as the Atlantic fit, and it was possible to generate several alternatives using the depth data from submarine continental margins on their own (Figure 2.3).

Some extra information was needed that would enable geoscientists to discriminate between these alternatives. The most useful proved to be the geological record on each of the component continents.

The geological match between continents

Jigsaws are fun because we must match both the shape and the pattern of adjoining pieces. To identify once-joined continents, we must match not only the continental shape but also the geological pattern by matching geological features across the two components. We might extend this line of reasoning further to surmise that just as jigsaws are built of groups of pieces that contain a recognizable pattern, so configurations of continents can be recognized by their geological continuity.

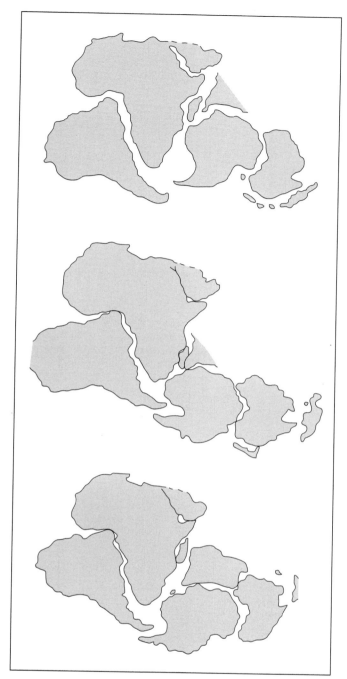

Figure 2.3   There are several ways of fitting the continents together based on the geometry of the continental margins alone. This Figure shows three different possibilities, indicating that extra information is needed to discriminate between them.

One of the most convincing geological fits is that of Africa and South America based on the outcrop pattern of their basement rocks. Figure 2.4 shows two suites of Precambrian basement rocks, late Precambrian rocks dated at 600–2 000 Ma and Archaean cratons older than 2 000 Ma, that outcrop on both sides of the South Atlantic. The two continents can be assembled, in this instance on the basis of Bullard's geometric fit of the continental margins, so that there is a superb correlation between these geological units on both continents. Looking at the complex detail of interfingering of the two rock groups on Figure 2.4 leaves no reasonable doubt that the two continents were once joined.

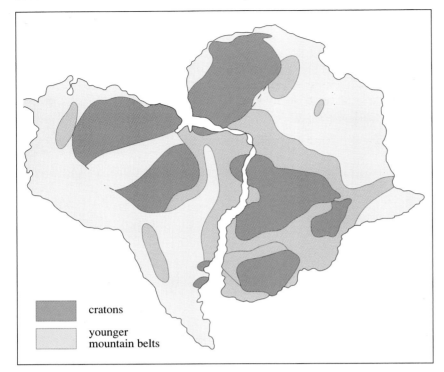

Figure 2.4   Correlation of Precambrian rocks between South America and Africa. Both Archaean (older than 2 000 Ma) and late Precambrian (600–2 000 Ma) outcrops match exactly across the South Atlantic.

cratons

younger mountain belts

Matching rock lithologies across oceans locates continents that were formerly joined and strengthens the notion of continental drift. For our purposes in this Course, we want to quantify continental drift; as yet, all we know is how far apart the continents are at the present day. We would like to be able to say when drift started and at what speeds and directions the continents moved apart. The geological record can provide extra information that helps geoscientists start to quantify continental drift. Often, the type of climate in which rocks were deposited can be identified from study of the rock types or fossils within the rock sequence.

Two good examples are provided by glacial deposits and coal. Rocks of late Carboniferous to early Permian age (some 300 Ma old) can be found widely across the world. In Britain, continental Europe and the USA, rocks of this age are coal-bearing. We know, from a wealth of information about the type of plants that grew, the animals that lived in the swamp and the detail of sediments that were deposited at the time that they formed on flat coastal plains near the late Carboniferous equator. Northern Europe must have been sited in the tropics in late Carboniferous times.

ITQ 2

Late Carboniferous coalfields are found in Britain near Newcastle at 55° N. If these coals formed in the tropics (i.e. 23° S–23° N), what is the minimum distance Britain has drifted in 300 Ma? The radius of the Earth is 6 370 km. (For this purpose, assume that the Earth is a sphere and that it hasn't changed dimensions in the past 300 Ma.)

ITQ 3

From your answer to ITQ 2, what is the average minimum rate of northward drift of Britain in mm yr^{-1} since the late Carboniferous?

In Africa, India, Antarctica, Australia and South America in late Carboniferous times ice-sheets covered the land, and sediments with characteristic glacial features were deposited (Figure 2.5). The extent of the glacial deposits not only suggests that these continents were joined at the time but also that the supercontinent they formed lay in polar regions. The continent of India must have moved considerably from polar regions to its present position in the tropics at some time in the past 300 Ma. Repeating the process of ITQs 2 and 3, this represents a drift of some 9 500 km (65° S to 20° N) and a drift rate of at least 30 mm yr^{-1}, depending over what exact time period the drift took place.

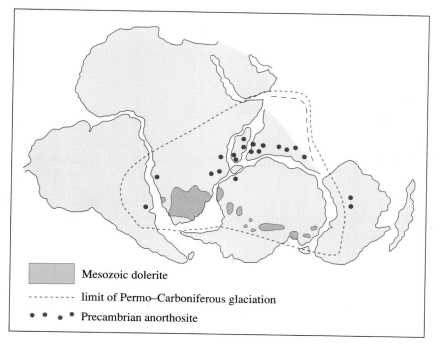

Mesozoic dolerite

-------- limit of Permo–Carboniferous glaciation

• • • • Precambrian anorthosite

Figure 2.5 Map showing correlation of marker climatic zones and distinctive rock types in continents that now border the Southern Ocean. Distinctive suites of igneous rocks (anorthosites and dolerites) and glacial deposits indicate that these continents were joined together for considerable parts of the geological past.

However, in our eagerness to quantify, we must be careful not to elevate assumptions to the status of facts. Glacial deposits might indicate polar latitudes, but equally they might have formed under an extensive mountain ice-sheet. Much of the Himalayas are ice-covered at the present time although they lie in sub-tropical latitudes. Has India moved half-way across the globe or has it simply subsided? Geoscientists need to be able to prove latitudes in a way that is independent of climatic belts. The Earth's magnetic field provides the answer.

Palaeomagnetism

The Earth has a magnetic field derived by convection of the electrically conducting liquid outer core. At the Earth's surface, the magnetic field is similar to that produced by a short bar magnet with poles deep inside the Earth and its axis roughly aligned parallel to the Earth's rotation axis. The magnetic field can be visualized by means of lines of force, which are parallel to the Earth's surface at the magnetic equator and dip steeply at high latitudes. In olden days, navigators could tell roughly at what latitude they were by using this feature of **magnetic inclination** and measuring the angle at which a compass needle dips.

Some rocks, particularly basalts, contain natural magnets in the form of the iron oxide mineral, magnetite. Magnetite crystals become magnetized in the direction of the prevailing magnetic field at the time of their formation. For example, young basalt samples from latitudes near the magnetic poles show steep inclinations within magnetite crystals, while similar samples from equatorial regions show shallow inclinations.

Magnetite-bearing rocks can therefore be used to indicate the latitude that the site of the sample occupied at the time it formed. (As the Earth's magnetic field is axially symmetric, magnets are no use as indicators of longitude.) We say that they form good **palaeolatitude** indicators. Basalt contains a significant proportion of magnetite crystals and is thus an excellent recorder of palaeolatitude. When the basalt erupts, magnetite crystals become aligned according to the strength of the magnetic field and the latitude of the eruption. As the lava cools, the mini-magnets become frozen into the rock, aligned in the prevailing magnetic field. Samples of basalt with a known present-day orientation can be collected and examined in the laboratory to give information about the magnetic field orientation at the time the sample cooled.

If a large expanse of basalt can be sampled, palaeomagnetic readings from well-separated localities can be used both to investigate the latitude at the time of the eruption and how much rotation the area has experienced. Magnetite-bearing rocks therefore are able to provide a great deal of data concerning plate movements.

ITQ 4

In general terms, how can palaeolatitude readings be used to estimate plate rotations?

This palaeomagnetic approach can be used to solve our problem of whether India drifted from a polar location to its present tropical one. Parts of western India are covered by a vast sheet of basalts some 65 Ma old. These basalts give magnetic palaeolatitude readings of between 30° and 35° S. India must therefore have drifted northwards for over 5 000 km in the past 65 Ma, and before that (300 Ma ago) it must have been located in polar regions.

ITQ 5

Is this last sentence accurate?

ITQ 6

Assume that India did lie south of 65° S some 300 Ma ago. Calculate the minimum northward drift rate for India over the periods (a) 0–65 Ma and (b) 65–300 Ma. Express both in $mm\,yr^{-1}$. Which is the higher?

Apparent polar wander

The corollary of measuring palaeolatitudes from magnetite-bearing samples is that such samples can also be used to locate the fossil magnetic poles (**palaeopoles**), as long as we know the palaeolatitudes of two or more samples of exactly the same age. In fact, palaeopole location and palaeolatitude calculation are really just two aspects of the same measurements. We can, however, present what is essentially the same data set in two ways (Figure 2.6).

Figure 2.6a shows palaeolatitude variations for the South American continent plotted on a view of the Southern Hemisphere. This view assumes that the magnetic pole is fixed and records the movement of the continent implied by making that assumption. Palaeolatitude readings tell us that this view of continental drift approaches the true picture. Figure 2.6b gives us a different approach to the same problem. Here, the continent is held in a fixed position and the apparent position of one of the poles is plotted. Because in this diagram the pole itself appears to drift, this approach gives us an **apparent polar wandering curve** for the continent in question.

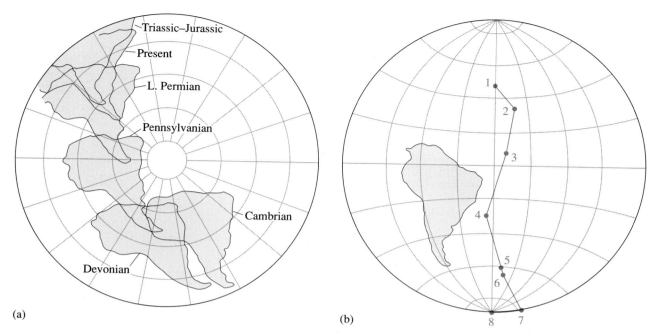

(a)   (b)

What is the point of this approach if we know that it's the continents that drift, not the poles? The answer to this question has two parts. First, apparent polar wandering curves can be plotted for many places on the same continent, as a verification of the palaeolatitude data. The data set does not have to be from samples of the same ages at all sites. The second use of apparent polar wanderings is to identify whether and when continents were joined and when they drifted apart. This use of apparent polar wandering curves is particularly valuable for our purposes since it provides verifiable data that prove the continents drift apart. This is exactly what we need to demonstrate plate movement.

Sea-floor spreading, magnetic reversals and the dawn of plate tectonics

Establishing that continents drift poses a further question: how is this drift achieved? As continents move apart, surely they must leave a 'gap' in the site they once occupied? This gap is now filled by ocean, so does oceanic crust form to fill in the space left when continents drift apart?

This idea is known as **sea-floor spreading**, and was proposed in 1961 by R. S. Dietz. He suggested that new ocean lithosphere is created by upwelling and partial melting of material from the underlying asthenosphere. As new oceanic crust is created, the ocean would grow wider and the continents at its margin would therefore move apart. Dietz proposed that mid-ocean ridges were the sites of generation of new oceanic crust.

There is another use of palaeomagnetism that introduces a completely new and exciting line of evidence for investigating plate movements, and was able to test the theory of sea-floor spreading effectively. For decades, geoscientists had known of the geological evidence supporting the theory that the continents move. Since the 1950s, some of them concentrated their studies on the oceanic crust that separates the continents to discover more about the process of drift.

ITQ 7

Dietz proposed the theory of sea-floor spreading. Suggest how he could test it, using the age of oceanic crust.

Figure 2.6 Two methods of displaying palaeomagnetic data. Diagram (a) assumes the magnetic poles are fixed over time, and records the apparent latitude drift of a continent. Several readings are needed to constrain longitude. Diagram (b) assumes that the continent has remained fixed over time, and records the apparent wandering path of the pole. No longitude assumptions are needed here either, as long as at least two samples of the same age have been measured. *Note:* Pennsylvanian = Carboniferous.

Then, as now, it was prohibitively expensive and time-consuming to collect sea-bed samples on the scale necessary to make a detailed grid of palaeomagnetic readings of the ocean floor; a remote-sensing method was clearly more appropriate. It was comparatively easy to measure the intensity of the Earth's present magnetic field using magnetometers towed behind ships. From the mid-1950s, magnetic surveys were carried out both in specific sites of interest, and routinely during the voyage from one site to the next. A substantial database of the intensity of the present-day magnetic field was quickly built up.

The Earth's magnetic field is affected by time-of-day variations, by sunspots and by regional magnetic variations due to anomalies in that part of the Earth's core. When corrections for these variables are made, the **magnetic anomaly** that results depends entirely on the magnetic properties of the crustal rocks in the survey site. This in turn depends largely on the mineralogy of those rocks.

We know (as did these geoscientists) that the upper part of the ocean floor is made up of a thin layer of sediment that overlies basalts. The sediment cover might be expected to be comparatively poor in magnetic minerals and thus to have little effect on the Earth's magnetic field. The basalt layer beneath should be rich in magnetic minerals and would be expected to produce a significant local magnetic anomaly. But basalt is basalt (or so it was thought!), and the investigators were expecting uniform magnetic properties across the oceans.

What they discovered excited them as much as it surprised them. The magnetic field was anything but uniform, instead containing an unexpected pattern of linear magnetic 'stripes'. These defined long, linear zones of rapid anomaly change, bordering an area of relatively uniform positive anomaly on one side and an area of relatively uniform negative anomaly on the other. Early discoveries off the western seaboard of the USA were quickly confirmed in other areas of ocean floor. They also found that the stripes were quite extensive laterally, and that they could trace individual anomalies for hundreds of kilometres along the ocean floor, until they stopped at prominent fault-like zones (Figure 2.7).

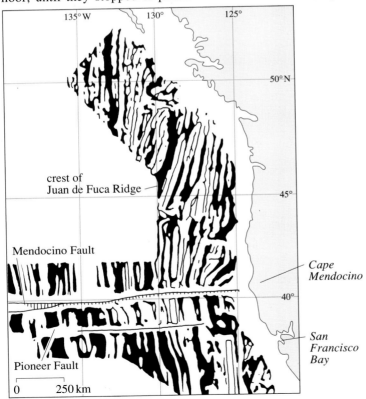

Figure 2.7 The pattern of linear magnetic anomalies discovered on the ocean floor. The black-and-white stripes represent zones of normal and reverse magnetization within sea-floor basalts.

Further detailed studies revealed two other interesting facts. The pattern of anomalies did not actually stop at these cross-faults but appeared to be displaced, sometimes by distances of over 1 000 km. The anomaly pattern also appeared to be symmetrical across the mid-ocean ridges. Particularly wide zones of uniform anomaly recognizable on one side of the ridge usually had a counterpart at about the same distance on the opposite side of the ridge.

ITQ 8

From your studies of Block 1, suggest three possible causes of these linear magnetic anomalies.

Seismic reflection and refraction data and dredge samples eliminated both compositional variations in the ocean-floor basalts and highly variable types and thicknesses of sediment cover in a relatively small area as the causes of the stripes. Therefore, stripes had to represent variations in the amount or direction (or both) of the Earth's magnetic field. They could be explained completely if the magnetic field had been subjected to frequent pole reversals over time. Each positive anomaly might represent a time during which a magnetic field of one polarity was established. When the polarity changed, the field would rapidly decay, to become re-established shortly afterwards with the opposite polarity, giving a corresponding stripe of negative anomaly. The boundary zones of a rapidly changing anomaly represented the time of decay of the magnetic field and its re-establishment with new polarity. Surprisingly, this elegant idea was not new; geomagnetic field reversals had been invoked early in the 20th century to explain reversed magnetizations in some land-based volcanic rocks.

When it became clear that the ocean-floor record must represent magnetic field reversals, a further problem presented itself. Magnetic reversals in *time* would be understandable, but how could magnetic reversals in *space* be explained? Why should the field reverse again and again across the ocean floor?

In 1963 (in a paper in the Journal *Nature*), two British geoscientists, F. J. Vine and D. H. Matthews, proposed an hypothesis which elegantly explained magnetic reversal stripes, linked these to sea-floor spreading and laid the foundation for plate tectonics. They suggested that new oceanic crust was formed by the solidification of basalt magma extruded at mid-ocean ridges. As it cooled, the magma acquired a magnetization in the same orientation as the prevailing global magnetic field. As the forces that drive sea-floor spreading continued to operate, the ocean margins would move apart and more oceanic crust would be generated at the ocean ridge. If the magnetic field reversed, then newly erupted basalt would become magnetized in the new ambient magnetic field which would be significantly different from the pre-existing one. Not only would the new oceanic crust show a different magnetic polarity from its neighbour, but it would also be slightly younger. As new oceanic crust continued to form, so stripes of basalt of the same age and magnetic polarity would spread away from the ridge. Ocean crust can therefore be considered as the 'tape' in a global 'tape recorder', in which the reversal history of the Earth's magnetic field was constantly recorded. Reading outwards one way from mid-ocean ridges should give a record of reversals over time which could be matched with the record read in the other direction.

The Vine and Matthews hypothesis gave a very effective tool for comparing plate movements. If the magnetic record, in terms of sequence of reversals, could be established in oceanic crust between just one ridge and its adjacent continent, this sequence could be compared with the other half of the same ocean to see if spreading from the mid-ocean ridge was symmetric. The record could also be compared with other oceans to see if spreading took place there at the same time, and also whether spreading in that ocean was symmetric or not.

The magnetic record, read away from a mid-ocean ridge as a known sequence of magnetic polarity switches, would establish a **magnetic time-scale**, which could then be used to compare the relative ages of oceans near to spreading ridges (Figure 2.8).

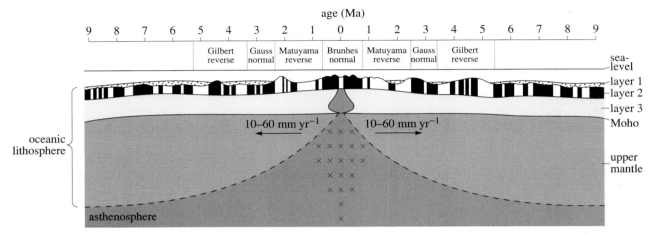

If samples from known places in the magnetic time-scale were dated independently, say by radiometric dating of suitable samples, then the actual age of components of the magnetic time-scale in millions of years could be determined. More interestingly, if a sample of known distance from a ridge could be dated, a rate for sea-floor spreading could be calculated.

Figure 2.8 Magnetic stripes can be investigated on either side of a mid-ocean ridge. By comparing and dating zones of normal and reverse magnetization, a time-scale for magnetic reversals can be established.

Implications of sea-floor spreading for plate boundaries

Before we move on to estimate rates of continental drift and sea-floor spreading, we should consider the implications of sea-floor spreading at mid-ocean ridges and the nature of plate boundaries. If oceanic crust is being formed at mid-ocean ridges and the margins of the ocean are moving apart, we can consider the mid-ocean ridge to be a *constructive* feature. As these ridges are also plate boundaries, we can refer to them more generally as **constructive plate boundaries**. This is altogether more useful than mid-ocean ridges, because many of the constructive boundaries do not lie in the middle of oceans. For example, the East Pacific Rise is far from being centrally sited in the Pacific Ocean. Confusingly, many ridges that do lie near the middle of oceans are not constructive plate boundaries. Good examples are provided by the Hawaii 'Ridge' in the Pacific Ocean and the Ninety-East Ridge in the Indian Ocean. We shall come across other synonyms for this type of boundary later in this Block, but we will try to use the formal name **constructive boundary** when we refer specifically to this type of plate boundary.

The Earth is neither expanding nor contracting in overall terms, and this fact has some profound implications on other types of plate boundary. If substantial amounts of new oceanic crust are being generated on a frequent basis at constructive boundaries, and if the overall volume of the Earth is not increasing, then oceanic crust must be actively destroyed at some part (or parts) of the globe, in equal measure to the amount being generated. **Destructive plate boundaries** are therefore an integral part of the theory of magnetic reversals, sea-floor spreading and the Vine and Matthews hypothesis. The existence of constructive plate boundaries requires the existence of destructive plate boundaries to maintain a constant-volume

Earth. We shall consider the exact nature of these boundaries in detail later.

Lastly, what about the oceanic fault-like structures that seem to displace the magnetic stripes? Are these destructive boundaries? Again, we shall consider these in detail at an appropriate point in the Block, but we can make an analogue model that helps us to deduce the nature of these structures.

Consider a tablecloth, one that just fits a smooth table exactly, and imagine tearing it in half down the middle. If you were to move the torn tablecloth apart in the middle of the table, it would droop down at *two* of the table edges (Figure 2.9).

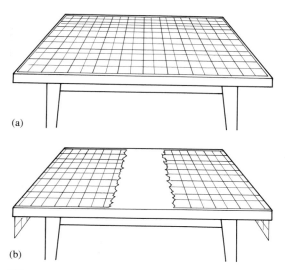

(a)

(b)

Figure 2.9 Implications of spreading axes. In (a), the tablecloth fits the table exactly. If it is torn and moved apart (b), then not only is there 'excess' tablecloth at the sides of the table but the tablecloth also moves laterally past the top and bottom of the table. Something similar happens when plates spread apart!

We can consider the ripped part of the cloth to be a constructive boundary (where new cloth needs to be created!) and the two drooping edges to be destructive boundaries (where there is 'excess' cloth to be destroyed). At the other two table edges, though, the cloth is neither being 'created' nor 'destroyed'; it is simply moving parallel to the table edge. This is a type of boundary (it certainly *is* a cloth boundary) but it doesn't fall into either of our existing categories for plate boundaries.

An analogous situation exists with plates, where they move past each other but are neither being created nor destroyed. This is in effect what is happening at the fractures which displace the magnetic stripes. As these boundaries do not involve making or destroying oceanic crust, they are termed **conservative plate boundaries** although we might note that the plates experience a liberal amount of labour in getting past each other! There are some excellent examples of conservative plate boundaries in the Pacific Ocean, at 130° W 55° S and between the words 'Pacific Plate' and 'Antarctic Plate' on that part of the Smithsonian Map.

2.2.1 RELATIVE PLATE MOVEMENTS

Let us now look at the ways in which relative motion can be estimated at plate boundaries. Spreading occurs away from an ocean ridge at constructive boundaries, so it is possible to refer either to the movement of one plate away from the other plate, or to the movement of one plate away from the ridge. We call the former the **spreading rate**, and the latter the **half-spreading rate**. This terminology makes the assumption that the rate of spreading is equal on either side of the ridge; we shall look at that assumption in more detail later. Both these are *relative* rates of movement, since we are measuring the movement of one side of a constructive boundary *relative* to the other, or to the ridge.

Relative rates of movement can be estimated on two time-scales: the **average rate of movement** is estimated over a period of geological time (generally exceeding a few Ma), and the so-called **instantaneous rate of movement** is estimated over much shorter time periods.

Using magnetic reversals

A simple method for estimating the average rate of sea-floor spreading at ocean ridges can be developed by comparing the pattern of linear magnetic anomalies on the ocean floor with the magnetic polarity time-scale. This method is best applied to oceans that have formed during periods measured in tens of millions of years.

ITQ 9

Using the data shown on Figure 2.10, estimate the average rate of relative spreading in the South Atlantic from 75 Ma to the present day. Express your answer in mm yr^{-1}. Is this average a spreading rate or a half-spreading rate?

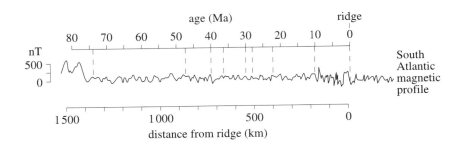

Figure 2.10 Magnetic anomaly profile across and perpendicular to the western flanks of the Mid-Atlantic Ridge. The broken lines indicate the ages of particular magnetic reversals.

This method can be used to provide good estimates of the rate of movement of ocean floor away from constructive boundaries. Where ocean-floor magnetic anomalies of known age can be identified at a measured distance apart and on opposite sides of a conservative plate boundary, they can also be used to determine a rate of relative motion between two plates, using the same technique. However, this method cannot be used to estimate relative rates of plate motion at a destructive plate boundary.

ITQ 10

Why not?

Using ocean-floor bathymetry

Ocean ridges have an average width of around 1 000 km and rise above the surrounding abyssal plains by between 2 and 3 km. The ocean floor is therefore higher at constructive boundaries than it is in the centre of the ocean plates. A simple explanation for this topographic difference might be that hot magma erupted at ocean ridges cools, contracts and becomes denser with time. The ocean floor therefore 'sinks' as it moves away from the ridge.

It follows that depth to ocean floor can be used to give a rough age of the oceanic crust itself. This is shown graphically in Figure 2.11. Furthermore, if the distance from the spreading ridge and the depth to a particular piece of ocean floor are both known, then the average rate of plate motion relative to the ridge can be calculated.

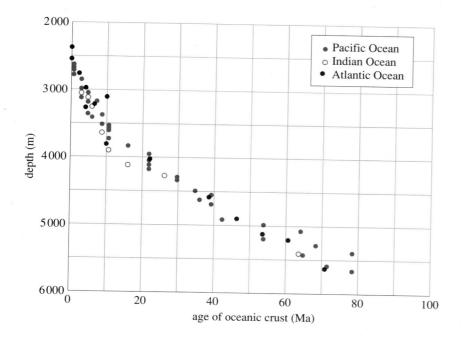

Figure 2.11   Observed relationships between depth to the ocean floor (both ocean ridges and abyssal plains) and age of formation of oceanic crust for the Pacific, Indian and Atlantic Oceans.

ITQ 11

Look on the Smithsonian Map at the spreading ridge between the Pacific and Antarctic Plates at 115° W 50° S. How far is the 4 000 m depth contour from the spreading ridge in the Pacific Plate? Using the data in Figure 2.11, how old must the oceanic crust be here? What therefore is the relative half-spreading rate (in mm yr^{-1}) of the Pacific Plate away from that ridge? What is the relative half-spreading rate of the Antarctic Plate here?

2.2.2   TOWARDS TRUE PLATE MOTIONS

So far, we have been concerned with the relative rates of motion between plates. Rather unfortunately, this does not actually tell us anything about the actual motion of plates with respect to some fixed coordinates within the Earth. Just because one plate appears to be moving relative to another, we cannot be sure whether both plates are really moving or whether one is actually stationary and the other plate is moving. You will no doubt have thought about problems like this if you have ever travelled on a train. Standing on the station platform, the train clearly moves out of the station, but for the observer actually sitting on the train looking out of the window, it is the station that appears to be moving. Common sense tells the traveller that it is actually the train that is moving.

When it comes to determining true plate movements (often confusingly called **absolute plate motion**), things are not so simple. Standing on one plate and looking at another, we cannot tell whether we are moving and it is stationary, or the other way round. Perhaps both are moving. Continental drift and magnetic stripes tell us that some type of movement is taking place, but only in relative terms. Try ITQ 12 to appreciate the problem.

ITQ 12

Consider the situation shown in Figure 2.12. This shows a section through the Earth from the Atlantic Ocean, through Africa to the Indian Ocean. The average relative spreading rates for the Mid-Atlantic Ridge and the Carlsberg Ridge are shown. For the situations tabulated below, estimate (i) the relative rates and (ii) the directions of plate motion for the African Plate.

Table 2.1   For use with ITQ 12.

	(i) Relative rate of plate motion	(ii) Direction of plate motion
(a) American Plate fixed		
(b) Australian Plate fixed		

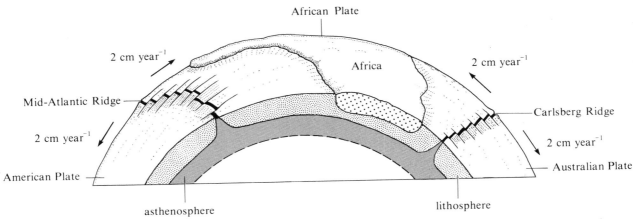

This ITQ brings up two important points about plate movement. First, measured directions and rates of plate motion must be relative one to another, and secondly, plates cannot be fixed in relation to the asthenosphere below. This second point follows from considering the change in size of plates. As the African Plate has grown in size, at least one of its plate boundaries must have moved outwards, away from their initial site relative to the asthenosphere.

How can we tell which plates and plate boundaries actually move? What is needed is some fixed set of features (like the station in the train example above) which we can use as a **fixed frame of reference**, against which to measure relative movements. This should give us the true movement direction and speed of each of the Earth's plates, which we shall call **true plate motion**.

There are in fact two problems here; how can we find a reference frame to determine the true motion of the Earth's plates, and how can we tell if our 'fixed' reference frame is in fact fixed? Does it actually move about, too? If so, true movements aren't!

Of course, we can somewhat arbitrarily *assume* that a certain plate or plate boundary is stationary and therefore has zero true motion. The true motions of other plates can now be calculated using observed relative motions of this fixed plate with adjacent plates (that was what you did in ITQ 12). In doing this we have necessarily made an arbitrary decision about which plate to consider fixed, and made an assumption which is almost certainly unjustified. So why bother? The value of this method is that it allows us to calculate true plate motions for all plates if we can once determine the true motion of just one plate (in some way other than guessing!).

Figure 2.12   Section through the Earth from the Atlantic Ocean through Africa to the Pacific Ocean showing the average relative spreading rates for the Mid-Atlantic and Carlsberg Ridges.

Using hot spots as a reference frame

Some isolated volcanic areas, such as Hawaii, and the Canary Islands off the northwest African coast, are distant from either constructive or destructive plate boundaries. These volcanic areas, together with unusually active volcanic areas that do lie on constructive margins, such as Iceland, the Azores and the Galápagos Islands, are termed **hot spots**. The origins of hot spots are discussed in Block 3. For our purposes it is sufficient to note that in several instances, particularly in the Pacific Ocean, active hot spots lie at the end of linear chains of volcanic islands.

Figure 2.13 shows three chains of volcanic islands and seamounts in the Pacific Ocean. The volcanoes that are active at present are located at the southeastern extremity of each chain, and the ages so far determined suggest that the volcanic rocks become progressively older in passing from southeast to northwest along each of these chains.

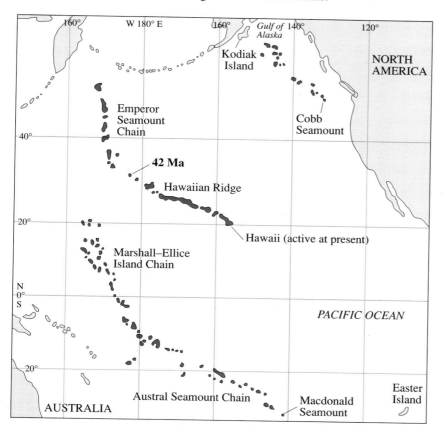

Figure 2.13 Three seamount and island chains in the Pacific Ocean: Hawaiian Ridge–Emperor Seamount Chain, Marshall-Ellice Islands–Austral Seamount Chain and Kodiak Island–Cobb Seamount Chain. The youngest volcanoes, at the south-eastern end of each chain, are named.

We might deduce that such volcanic chains were produced not because the source of heat needed to generate the magma was moving, but because the Pacific Plate itself was moving. A narrow heat source consisting of material flowing upwards from deep in the mantle might therefore be fixed in relation to the moving lithosphere above it (Figure 2.14).

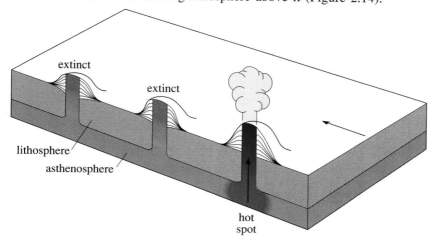

Figure 2.14 Stylized block diagram (not to scale) illustrating how a volcanic-island chain could be formed by an oceanic plate moving over a stationary hot spot. In reality, the individual volcanic islands will be projections in a ridge rather than separate structures. The ages of the islands increase towards the left; new islands will appear on the right as the motion continues. The curved lines through the volcanic islands symbolize the products of successive eruptions from the volcanoes.

ITQ 13

From the age of the volcanic island shown on the Hawaiian Ridge in Figure 2.13 and from the Smithsonian Map, deduce the average true rate and direction of movement for the Pacific Plate over a hot spot assumed to be fixed. (Distances may be determined by noting that 1° of longitude is equivalent to approximately 100 km.)

In this example, the average rate of true motion of the Pacific Plate throughout much of the Tertiary is about $70\,\mathrm{mm\,yr^{-1}}$, relative to the Hawaiian hot spot. Having established the true velocity of one plate, the true motions of the other plates can now be calculated from their motion relative to the Pacific Plate using the method outlined above. In reality, the type of calculation performed in ITQ 13 has been repeated for all the known hot spots, to provide a close check on the assumptions of relative plate movement made from magnetic stripes.

Are hot spots stationary?

Identifying the Hawaiian hot spot as a stationary feature detached from the moving Pacific Plate is critical in calculating the true movement amounts and directions of the Pacific Plate. As we have just seen, we need only to know the true movement of one plate to be able to calculate the true movements of the others. But there is a potential flaw in this argument. How do we know that the Hawaiian hot spot *is* stationary? If the hot spot itself has actually moved its location within the Earth over the past 45 Ma, then our calculations are incorrect and we no longer know the true velocity of the Pacific Plate. By extension, we cannot be sure about any of the other true movements either! It has become critical to our argument to find out if the Hawaiian hot spot is actually stationary or is itself moving.

We can approach this using data from other hot spots. If we can recognize another hot spot on the Pacific Plate which has a chain of volcanic islands like Hawaii, and if we can calculate the velocity of the Pacific Plate over the past 45 Ma from this new information, we can compare the two **movement vectors** (a vector is a quantity which has both magnitude and direction) for the Pacific Plate. If the two answers are the same, then either the two hot spots have been stationary over the past 45 Ma or they have both moved relative to the Pacific Plate by exactly the same amount and direction over that time. If the two answers are different, then one hot spot has definitely moved relative to the other, and we cannot be sure which one is actually moving (the train-and-station problem again). Of course, if we can find three, four, five or more hot spots within the Pacific Plate and they all give the same movement vector, we can become more confident that the hot spots are stationary and that we know the true motion of the plate.

In 1972, an American geoscientist, W. J. Morgan, noted the similarity of the orientation of chains of major volcanic islands and seamounts on the Pacific Plate. He noticed that the geometry of the Hawaiian–Emperor and the Austral–Marshall Island seamount chains shown on Figure 2.13 can be matched to that expected by moving a rigid Pacific Plate over the hot spots fixed at the present locations of Hawaii and the Macdonald seamount, respectively. Subsequent studies have extended that early work by incorporating the distribution of the ages of the volcanoes as well as the geometry of the volcanic chains, and these confirmed the fixed relative positions of two of these Pacific hot spots over the past 45 Ma.

The coincidence of two hot spots giving the same result for the Pacific Plate strictly means that these hot spots have not moved relative to each other. We can be rather more confident than we were from just the Hawaiian example that hot spots really are stationary, but we cannot yet eliminate the possibility that they are coupled together in some way, and drifting as a constellation of hot spots. Obviously, the more hot spots that give the same result, the more likely it is that the hot-spot system is stationary within the Earth.

❑   Is this logic sound?

■   Not strictly. We can link all the hot spots together and know for certain that they don't move relative to each other, but we still can't say that they are not drifting around the globe as a linked set. It's a superb example of the frailty of deductive reasoning — it sounds plausible, but it is in fact shaky.

There is a limit to the number of hot spots that underlie the Pacific Plate, but we can extend the data set if we include hot spots from adjacent plates separated from the Pacific Plate by constructive margins, using the magnetic stripes generated at the common ridge to predict true movements.

If we use the assumed true motion of the Pacific Plate for the past 10 Ma (based on the Pacific hot spots) and the known relative movement direction and amount between the Pacific Plate and an adjacent plate (from the magnetic stripes formed during the past 10 Ma), we can calculate the true movement of the adjacent plate over that time. This is an extension of the method used in ITQ 12.

We then plot the position of a present-day seamount on the adjacent plate which is believed to overlie a hot spot. This gives the position of the present-day hot spot. Now we can use the true movement vector for the adjacent plate (which we have just worked out) to predict the site of a seamount or volcano that would have existed on the adjacent plate above the hot spot 10 Ma ago. This gives us the average track of the adjacent plate over the past 10 Ma. We can continue the process as far back in time as the dating of magnetic stripes and volcanic islands will allow.

We can perform this calculation for all known hot spots on plates adjacent to the Pacific Plate and separated from it by a constructive plate margin. A quick glance at the Smithsonian Map shows that plates that meet those requirements are the Nazca, Cocos and Antarctic Plates. This will give us, for each of those plates, a suite of predicted seamount chains that we can compare with the actual chains, and also will give us some predicted dates of volcanism along the chains.

We can extend this reasoning to plates adjacent to those plates, as long as they are separated from them by a constructive margin, and this way we can also include the African, South American, North American, European, Arabian, Indian and Australasian Plates (in fact, almost all known plates). It is a difficult exercise to do, involving spherical geometry and careful consideration of relative movement vectors between plates. The answer published in 1991 by two geoscientists, R. A. Duncan of Oregon State University and M. A. Richards of the University of California at Berkeley, is presented in Figure 2.15.

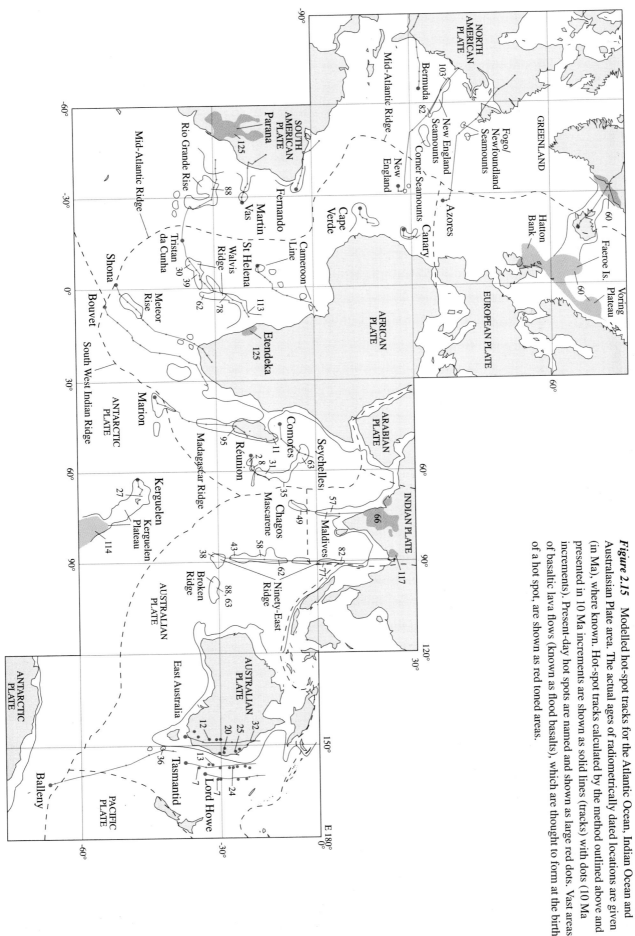

Figure 2.15   Modelled hot-spot tracks for the Atlantic Ocean, Indian Ocean and Australasian Plate area. The actual ages of radiometrically dated locations are given (in Ma), where known. Hot-spot tracks calculated by the method outlined above and presented in 10 Ma increments are shown as solid lines (tracks) with dots (10 Ma increments). Present-day hot spots are named and shown as large red dots. Vast areas of basaltic lava flows (known as flood basalts), which are thought to form at the birth of a hot spot, are shown as red toned areas.

Spend some time looking at Figure 2.15, as it shows many interesting features of hot spots. First, note the almost perfect match between the predicted lines of seamounts and volcanic islands (solid lines) and the actual named seamount chains. Associated with this there is an equally perfect match between the age of volcanism predicted by the procedure described above (dots every 10 Ma along the solid lines) and the known age of volcanism on the seamount chains. This fit is extremely good for Earth science data, and shows that the features you would expect to have developed in association with a fixed set of hot spots can actually be observed for almost every hot spot on the globe.

Secondly, note how the hot-spot chains on the same plate track in parallel with each other. Look at the ridges and seamount chains that come from the hot spots of Bouvet, Shona, Tristan da Cunha, St Helena, Cape Verde and Canary on the western side of the African Plate. These hot-spot trails all have exactly the same form. Compare for example the Shona trail with the Tristan trail. Slight variations in plate movement are recorded at 100, 80, 68, 50 and 40 Ma on the Walvis Ridge, which are duplicated exactly on the Meteor Rise and its extension towards South Africa.

Thirdly, note that the hot-spot trails themselves give a rough indication of the velocity (amount *and* direction) of individual plates. Look at the Ninety-East Ridge (understandably at 90° E in the Indian Ocean) and compare it with the Walvis Ridge west of Africa. These two span the same age range approximately (from about 115 Ma to 30 Ma) but the Ninety-East Ridge is almost twice the length of the Walvis Ridge. This tells us that over that period of the Earth's history the Indian Plate was moving northwards twice as fast (in absolute terms) as the African Plate was moving northeastward.

The fourth important observation from this map of hot-spot tracks is that hot spots may start life underneath one plate and finish up under another. This somewhat surprising observation is well shown by the Ninety-East Ridge. This hot spot started under India and Bangladesh some 117 Ma ago, and the Indo-Australian Plate (labelled as such on the Smithsonian Map) tracked northwards rapidly across it for some 80 Ma. Between 38 and 27 Ma, the constructive margin separating the Indo-Australian Plate from the Antarctic Plate must have crossed over the hot spot because hot-spot activity suddenly ceased on the Indo-Australian Plate and commenced on the Antarctic Plate. We can see that the Kerguelen hot spot appears to have been responsible for the Ninety-East Ridge and the flood basalts in Bangladesh. A similar story is told by the Azores and New England hot spots in the Atlantic Ocean.

The important point to grasp is that if hot spots are fixed in the Earth, and spreading ridges cross over them from time to time, then the spreading ridges themselves must move in real terms.

The magnetic reference frame and true polar wander

Earlier, we observed that hot spots must be fixed relative to each other, but noted that this did not necessarily mean that they were absolutely fixed as well. The hot spots could still be moving slowly as a linked set relative to the Earth. If the hot spots themselves are moving, we are not really any nearer to determining the true motion of plates. Furthermore, if hot spots move we cannot be sure that spreading ridges move *over* them rather than they *under* the ridges. How can we determine whether hot spots do move?

One way of doing this is to find some other way of calculating true plate movement and cross-check this with the true movements obtained by studying hot spots. As with the hot-spot reference frame, it will be enough to determine the true motion of any one plate; the others can be calculated from the relative motions of adjacent plates. Like we saw for hot spots, it should be possible to check any new method for determining true plate motion by predicting movements for one plate using data from another, and comparing the prediction with the measured values.

An independent method of calculating true plate motion in terms of latitude (but not longitude) is provided by the palaeomagnetic record. If we know where the palaeopole was at any time in question, we can work out the palaeolatitute of a particular group of rocks containing magnetic minerals. You have already seen that oceanic crust contains magnetic stripes; these can be used for palaeolatitude calculations as well as for dating relative motions.

First, the orientation of magnetic minerals is determined in an appropriate sample of known age on one plate. As basalts are particularly suitable, volcanic islands and seamounts are ideal sites. Secondly, the palaeopole is calculated from this sample. This can be checked by repeating the experiment for other sites on the same plate and of the same age. Thirdly, other samples from the same plate but of different ages are examined and their palaeopoles plotted. If the plate has moved relative to the magnetic field, the palaeopole will appear to 'wander' with time. The direction and amount of polar wander is in fact recording the true movement of the plate. Fourthly, the apparent polar wander path for an adjacent plate can be predicted for the same time interval by calculating the true movement for that plate from the relative motion of the two, and predicting the adjacent plate's polar wander path. This can be compared with the actual polar wander data for the adjacent plate.

This experiment has been done (in fact, it preceded the hot-spot experiment) and the true plate vectors produced are closely similar to those yielded by the hot-spot method. Frustratingly, we are still not out of the woods because the Earth's magnetic axis might itself wander with time. The same criticisms levelled against the hot-spot reference frame can be levelled against the magnetic reference frame! The problem does not in fact have a solution, because we can never be absolutely sure that any reference frame is stationary over time. We simply don't have independent data from the past.

We can, however, compare our two best methods. If we predict the polar wander path over a given time interval from the hot-spot reference frame and compare it with the measured polar wander curve from magnetic data, the two answers should be the same if both are absolutely stationary. Again, the answers will be identical if both move, but are fixed relative to each other. If the answers are different, one must move in relation to the other. The origins of hot spots are thought to be completely independent of the origin of the Earth's magnetic field, so if the two polar wander curves turn out to be identical we can be a little bit more certain that both hot spots and the Earth's magnetic field are stationary.

The results of such a comparison are presented in Figure 2.16 and you can see that although the two polar wander paths are very similar in shape and form, they are not identical. The difference between the two is termed **true polar wander** to distinguish it from the apparent polar wander of the Earth's magnetic axis relative to moving plates. The existence of true polar wander confirms that one reference frame must be moving relative to the other. The amount of this movement can be calculated by 'subtracting' the path given by one method from the path given by the other.

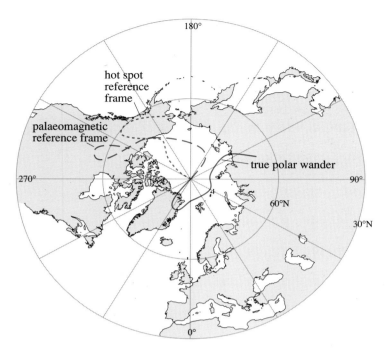

Figure 2.16 The polar wander path for the past 200 Ma deduced from palaeomagnetic studies (dashed line) compared with the polar wander path for the same time interval deduced from the hot-spot reference frame (dotted line). The palaeomagnetic path represents the apparent motion of the magnetic pole seen by an observer standing on the African Plate from 200 Ma to the present. The two paths are similar in overall pattern, but not identical. A true polar wander path can be calculated by subtracting one from another (solid line). This is thought to represent the movement of the Earth's core relative to the Earth's mantle caused by plate reconfigurations over that period. Points on each curve are at 20 Ma intervals.

Disappointingly, this still doesn't confirm the fixed nature of either reference frame, as either or both might be moving. We do know that the Earth's magnetic field is intimately linked to its spin axis and is generated in the core. Hot spots are derived from the mantle. This relative motion between reference frames must reflect relative motion between the core and the mantle.

After all this, where do we now stand? Well, the logic of the situation is clear. We will never be certain that either the hot spot or the magnetic frame has been constant throughout time until we invent a machine capable of time travel. We know that those two frames have moved slightly relative to each other. We also know that the plates move dramatically relative to both. We don't know and probably never will be certain that the hot-spot frame of reference is stationary, but it does represent the closest we will get to a stationary reference frame.

2.2.3 MEASURING MODERN PLATE MOVEMENTS

The rate of relative plate movements at the present day can be measured extremely accurately by using modern space technology. This, of course, gives us no insight into plate movements in the past.

Three independent methods are available to geoscientists:

Very long baseline interferometry (VLBI) makes use of radio signals from distant space. Quasars are the energetic cores of certain distant galaxies which produce enormous amounts of energy from a region no bigger than our Solar System. The signal from a particular quasar is recorded simultaneously by several radio telescopes, each many thousands of kilometres from the others. Signals arrive at different times, the magnitude of the delay depending on the distance between the recording stations and the direction each makes with the source. Typically, 10 or more quasars are each observed up to 15 times a day in any one experiment. This technique can measure changes in distance between the recording stations to within 20 mm.

Satellite laser ranging is similar in principle to seismic reflection profiling. The distance to an orbiting satellite or a reflector on the Moon's surface is calculated by timing the two-way travel time of a beam of laser light from two or more stations simultaneously. As lasers travel at the speed of light, the distance of the reflector or satellite from each measuring station can be simply calculated. This method can measure changes in distance between the recording stations to within 80 mm.

Global positioning by satellite (GPS) is a three-dimensional method which calculates the positions of stations around the globe from radio signals received at several specialist receivers. It is the most accurate and the most commonly used of the three methods, giving accuracies of approximately 1 mm in 10 km.

Applying these methods, the San Andreas Fault is moving on average at 30 mm yr^{-1} using laser rangefinding and at 25 ± 4 mm yr^{-1} using quasar interferometry. Britain is moving away from America at 19 ± 10 mm yr^{-1}, which is in close agreement with the figure of 23 mm yr^{-1} over the past 1 Ma, derived from magnetic stripe calculations.

For interest, and by way of a comparison, that's about the same rate at which your fingernails are growing!

2.2.4   HOW OLD ARE OCEANIC PLATES?

If we know rates of plate movement, then it is a straightforward matter to calculate the age of oceanic crust by measuring the distance from the appropriate mid-oceanic ridge. This provides a rather crude age estimate, and the technique can be refined by dating individual magnetic stripes using radiometric methods. Basalt samples are common across the oceans, so radiometric dates can be obtained from most sites when they are sampled. If samples of known polarity from known locations are dated, geoscientists can quickly build up a time-scale, based on the ages and locations of magnetic ocean-floor stripes. Together, these techniques enable us to tell the age of almost all oceanic crust across the world.

Look now at Plate 2.1. This map shows the ages of oceanic crust that forms the ocean floor across the Earth. A colour code indicates ocean-floor ages, based on dating of zones of magnetic stripes. Oceanic crust less than 20 Ma old is shown in reds; older crust is shown through a rainbow sequence of yellow, orange, brown, green and blue. Each colour change indicates progressively older rocks.

There is a distinctive colour change, from brown to green, at 80 Ma in this diagram. By looking at this change through the oceans, we can see at once that most oceanic crust across the globe is less than 80 Ma old. The whole floor of the eastern Pacific Ocean, most of the Indian Ocean and most of the Atlantic Ocean has been formed since then. In fact, areas of ocean floor older than 140 Ma occur only at the extreme edges of some parts of the ocean system. Oceanic crust over 160 Ma old occurs only in three areas: east of the USA and west of Saharan Africa in the Atlantic Ocean, and to the east of the Mariana Islands in the west Pacific Ocean. This latter site, the Pigafetta Basin, is the world's oldest sea-floor.

Dated magnetic stripes not only tell us the age of the oceans, but can also time the breakup of continents. The oldest oceanic crust that borders a continent must have formed at the start of sea-floor spreading, and therefore should record the age when that continent separated from its neighbour. For example, in the northern Atlantic, oceanic crust older than 140 Ma (blue colour on Plate 2.1) is restricted to eastern USA and western Saharan Africa. Separation of North America from this part of Africa must have commenced at this time. We can tell also that parts of the North Atlantic must have started to form before the South Atlantic, since the oldest oceanic crust that borders South America and sub-equatorial Africa is only about 120 Ma old. The following list gives some other ages of continental breakup:

Africa–North America	160 Ma ago
Africa–South America	120 Ma ago
Greenland–Scandinavia	80 Ma ago
India–Antarctica	80 Ma ago
Australia–Antarctica	65 Ma ago

At some continental margins, e.g. eastern South America, oceanic crust of essentially the same age is found along the whole length of the coastline. A complete sequence of magnetic stripes is preserved between coasts like these and the mid-ocean ridge. These are **passive margins** with no seismic and little volcanic activity.

At other margins, e.g. western South America, oceanic crust of different ages borders the continent, and many of the dated stripes from the western part of the Pacific Ocean are 'missing' in this eastern part. These coasts are **active margins** which are seismically and volcanically active.

ITQ 14

By contrasting the symmetry of dated magnetic stripes in the Atlantic and Pacific Oceans, what can we tell about the age and preservation of oceanic crust?

ITQ 15

Give two possible reasons for the 'missing' oceanic crust off the western coast of South America. Which is the more plausible?

Some of the most compelling evidence for plate tectonics is highlighted here, in the areas of 'missing' oceanic crust. If western South America was ever a passive margin, and it is known from palaeomagnetic studies that it was at least some sort of continental margin 200 Ma ago, then we must assume that oceanic crust between 200 Ma and about 70 Ma once bordered it. That oceanic crust is now missing; the site where it should be is now occupied by a tectonically active plate boundary. We can deduce that oceanic crust is being destroyed at some active continental margins.

SUMMARY OF SECTION 2.2

- The continents have drifted apart over geological time. We can show this by matching the continental margins of some adjacent continents and by tracing geological features across continents now separated by major oceans. Using these techniques, we can be confident that South America was once joined to Africa, for example.

- We can tell roughly how far some continents have drifted by identifying geological sequences of known age which contain distinctive climatic indicators. This way, we can infer that India moved from polar regions to its present position at an average rate of about 30 mm yr^{-1}.

- Some rocks (notably basalt) contain minerals that can record the Earth's magnetic field; these can be used as palaeolatitude indicators. Such measurements confirm that all the continents have changed palaeolatitude over geological history.

- Palaeomagnetism also shows the apparent wandering paths of the magnetic poles. We do not think that the poles themselves wander much, so the apparent wandering paths must be caused by continental drift.

- If continents move apart, then new ocean floor must be created to fill the gap.

- The ocean floor contains 'stripes' of basalt showing alternations of normal and reverse magnetic polarity. By dating ocean-floor basalts, a time-scale for magnetic reversals can be established. The magnetic polarity time-scale indicates that the youngest ocean-floor basalts occur near tectonically active mid-ocean ridges. These must be the sites of sea-floor construction.

- We can measure spreading rates at these constructive plate boundaries by using the known magnetic stripe time-scale. These measurements give us accurate information of how plates move with reference to each other.

- Comparison of calculated relative plate movements with those predicted from hot-spot 'drift' strongly suggest that the hot-spot frame is stationary. If so, it can then be used to calculate true plate movements. This observation is largely supported by comparison with the true wandering of the magnetic poles.

- Both plates and plate margins may ride over hot spots.

- Modern satellite and astronomical technology allows accurate measurements of the present-day plate movements. Measurements made using these techniques give plate motions closely comparable with measurements derived from magnetic stripe calculations.

- The oldest oceanic crust is only about 170 Ma old, and most oceanic crust is much younger than this. Oceanic crust older than this has been destroyed at tectonically active plate margins.

OBJECTIVES FOR SECTION 2.2

When you have completed this Section, you should be able to:

2.1   Recognize and use definitions and applications of each of the terms printed in bold.

2.2   Identify the geological and geophysical factors that provide evidence for plate motion.

2.3   Define the types of plate boundary which are an inevitable consequence of moving plates.

2.4   Use a variety of geophysical methods to distinguish between absolute and relative plate motion.

2.5   Estimate the age of oceanic crust, given both a magnetic stripe pattern and a magnetic time-scale for that particular area, or from oceanic water depth.

2.6   Estimate the relative spreading rates of two adjacent plates and describe the true movement pattern of any given plate from hot-spot tracks.

2.7   Describe several methods of measuring rates of present-day plate movement.

Apart from Objective 2.1, to which they all relate, the 15 ITQs in this Section test the Objectives as follows: ITQs 1, 2, 3, 4 and 6, Objective 2.2; ITQs 7, 8 and 9, Objectives 2.4 and 2.5; ITQs 11, 12 and 13, Objective 2.6; ITQs 14 and 15, Objective 2.3.

You should now do the following SAQs, which test other aspects of the Objectives.

SAQS FOR SECTION 2.2

SAQ 1 (*Objectives 2.1, 2.2 and 2.4–2.7*)

The theory of plate tectonics requires lithospheric plates to move constantly relative to each other and relative to some fixed global reference frame.

(a)   What geoscientific evidence tells us that plates are moving at the present day?

(b)   Which geophysical and geochemical data tell us that plates have moved in the past?

SAQ 2 (*Objectives 2.1, 2.3 and 2.4*)

Why do ocean-floor basalts contain magnetic 'stripes'? Why might geoscientists want to study these stripes?

SAQ 3 (*Objectives 2.1, 2.2, 2.4 and 2.6*)

Why do geoscientists believe the hot-spot reference frame is almost stationary within the Earth? What information does the hot-spot reference frame give us about plate movements?

2.3 OCEAN PLATE BOUNDARIES

We have seen that plates move relative to each other and to the Earth as a whole. If plates are rigid and are moving relative to each other, then three categories of plate boundary must exist.

(i) There must be places where the boundaries of two plates are *divergent*, or moving away from each other. We have already recognized that this situation exists within oceans. We know from the age of ocean-floor basalts that at these oceanic sites new ocean floor is being created. An appropriate name for these boundaries is therefore **constructive plate boundaries**.

(ii) If the overall surface area of the Earth is not increasing as a result of constructive plate boundaries, then there must also be places where ocean floor is being destroyed. These are called **destructive plate boundaries**. (These might also be places where two plates are convergent, or moving towards each other, although as yet we have not shown that this is the case.)

(iii) Plates need not always move towards or away from each other; they might simply slip sideways past each other. If they do this, then there is no reason why ocean floor should be either created or destroyed at this type of plate boundary. As these boundaries would conserve plate area, an appropriate name for these would be **conservative plate boundaries**.

(iv) For completeness, we should note that plate boundaries might be a combination of two of these basic types. It is unlikely that there would be any combinations of constructive and destructive boundaries, but combinations of conservative margins with both constructive and destructive margins could be expected and do in fact exist. An appropriate name for these types is **combined plate boundaries**.

This basic classification of plate boundaries forms another central theme in the theory of plate tectonics, and the concept is well shown on the schematic cross-section on the Smithsonian Map. You may find that you need to refer back to the Smithsonian cross-section at intervals as you read the following Section.

There is no reason why constructive, destructive and conservative plate boundaries should necessarily be restricted to oceanic crust. Indeed, they are not, but for simplicity we will deal first with the case of oceanic plate boundaries and extend the discussion in Section 2.5 to consider ocean–continent and continent–continent plate boundaries.

2.3.1 CONSTRUCTIVE PLATE BOUNDARIES

Constructive plate boundaries in the oceans are marked by long, relatively linear ocean ridges characterized by shallow earthquakes. They are sites where new ocean floor is actively forming. Ocean ridges are actually the longest tectonic features on the Earth's surface; the total length of the present-day ridge system is over 60 000 km. They are characteristically zones of tension. Because tectonic plates spread apart at ocean ridges, a constructive plate boundary has an alternative name of a **spreading ridge** or **spreading axis**, made up of several **spreading centres** strung out along the axis. More loosely constructive boundaries are also called **mid-ocean ridges**, but as we have already seen they are rarely placed exactly centrally within the oceans, and not all ocean ridges are constructive plate boundaries. Each of these terms has its own value, and we shall use each as and when it is appropriate. Their official name will remain 'constructive plate boundaries', though.

At spreading axes, plates move apart at rates varying from a slow 10 mm yr⁻¹ on parts of the Mid-Atlantic Ridge to a rapid 90 mm yr⁻¹ on parts of the East Pacific Rise. We shall start our study of constructive plate boundaries by looking first at the ridges themselves in some detail.

Spreading ridges and ridge topography

Spreading ridges themselves can be up to 3 km above their surrounding deep ocean plains. There is a gentle slope upwards from the deep ocean floor to the tops of spreading ridges; this slope is less than 0.5°, even for the steepest ocean ridges. Often the central part of the ridge takes the form of a prominent valley, called a **median rift**. The median rift can be 30–50 km wide and up to 3 km deep, although a figure of half that is perhaps more common. Median rifts do not occur on all parts of all spreading ridges.

Earthquakes are located at rather shallow depths on spreading ridges. The central parts of constructive boundary ridges are apparently the sites of relatively low-energy earthquake activity, compared with other types of plate boundary. (In places this may not be real but be due to poor sampling.) The rather low numbers of seismographic stations in ocean areas means that small earthquakes may often go undetected.

Large volcanoes are not a feature of ocean ridges, except where the ridge coincides with a site of hot-spot activity. Even in these cases, the volcanoes belong more strictly to the hot spot rather than to the ridge itself. Nevertheless, ridges are sites of high heat flow and are associated with frequent emissions of basaltic lava.

What is the significance of a median rift? Where exactly are median rifts developed, if they don't occur everywhere on the ridge system?

Figure 2.17 shows profiles across three parts of the present-day ridge network.

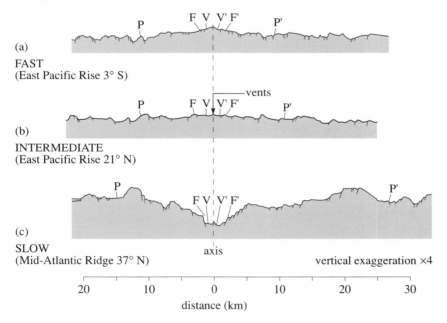

(a)
FAST
(East Pacific Rise 3° S)

(b)
INTERMEDIATE
(East Pacific Rise 21° N)

(c)
SLOW
(Mid-Atlantic Ridge 37° N)

axis

vertical exaggeration ×4

distance (km)

Figure 2.17 Cross-sections through three currently active parts of the modern ridge system, (a) across the East Pacific Rise at 3° S, (b) across the East Pacific Rise at 21° N and (c) across the Mid-Atlantic Ridge at 37° N. These sections are chosen to show the topography of fast-, intermediate- and slow-spreading ridges. The zone of currently active volcanic vents lies centrally on each ridge (V–V′). On either side of this there is a zone of fissures (F–F′), and outside this lies a zone of active faults (P–P′).

Figure 2.17 suggests that the exact shape of the central portion of ridges is controlled primarily by the spreading rate of the two plates. The East Pacific Rise at 3° S (Figure 2.17a) is spreading apart rapidly, at more than 90 mm yr⁻¹. This part of the ridge shows no median valley, and the zone of active faulting is restricted to close to the zone of active lava eruption.

By contrast, the Mid-Atlantic Ridge at 37°N (Figure 2.17c) is spreading slowly, at only 10–20mm yr⁻¹. It shows a deep median valley some 30km wide, which reaches depths of 2800m below the flanks. The central low portion is flanked by a wide zone of active faulting; steps down on these faults towards the crest generate the median valley.

The form of the median valley is not necessarily constant for any ridge, as Figure 2.17b shows. This is a section through the East Pacific Rise at 21°N, where spreading rates are intermediate between the other two. The form of the rift topography is different from the same ridge further south, with a slightly wider zone of active faulting and an evident if not pronounced central valley.

Figure 2.18 shows the Bouguer and free-air gravity anomalies across part of the Mid-Atlantic Ridge that has a pronounced median valley. You may remember from Block 1 that the free-air anomaly measures the actual gravitational pull from all the material beneath the recording station, whereas the Bouguer anomaly allows for density variations in the shallow subsurface rocks.

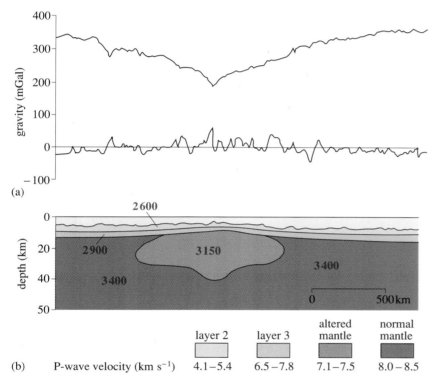

Figure 2.18   (a) Bouguer (top) and free-air (below) anomalies across part of the Mid-Atlantic Ridge. (b) One possible density model that could produce the observed anomalies, and which satisfies the seismic structure of Figure 1.58. Numbers in colour are modelled densities.

The free-air anomaly is relatively flat and near zero across the ridge. This tells us that there can be no mass excess or deficit beneath the sea-surface down to the level of compensation. The ocean ridge is therefore in overall isostatic equilibrium with the surrounding oceanic lithosphere. The Bouguer anomaly is strongly positive but with a local dip across the centre of the ridge. The positive anomaly occurs because of the raised levels in this part of the ocean floor (Section 1.8) but the ridge zone must contain material of lower density than the surrounding oceanic lithosphere, because it stands topographically higher than the surrounding ocean floor even though it's actually in isostatic equilibrium. The local dip in the Bouguer anomaly confirms that there is a relative mass deficit beneath the rock surface over the central part of the ridge.

The model in the lower part of Figure 2.18 is one of several explanations for the extent and density of the light material that exists beneath the ridge. Of course, if the density contrast were to be changed in the model, the area of anomalously light material would need to be altered also.

ITQ 16

What might this light material be?

How can these geophysical and topographic features be unified and used to explain what is happening at constructive plate margins? Before we can answer this question sensibly, we need to add to our data set the geological, geochemical and geophysical observations about the structure of oceanic crust that we introduced in Block 1. We can use seismic data and field observations made from onshore occurrences of oceanic crust (ophiolites) to tell us about the structure of oceanic crust.

ITQ 17

What is the structure of oceanic crust?

We want to be able to relate the composition of oceanic crust to the geophysical data and the topographic features we see at spreading ridges so that we can understand the processes that are taking place there. How can we do this?

ITQ 18

In other branches of science or technology, how might a theory or design be generated, and then tested? Would this process be different in, say, the arts or humanities?

ITQ 19

Which processes discussed in the answer to ITQ 18 would most help our understanding of the processes at spreading ridges?

An effective way of understanding ridge processes is by building an empirical model, using the technique of asking questions and finding answers. Erecting models from geological and geophysical observations and then testing them (logically or physically) is the exciting 'stuff of life' of scientific research. Discovering about the natural world by observation and deduction is what those who make a living from geo-scientific research (like the Course Team!) enjoy so much about their job.

We shall start our thought-modelling procedure by questioning why oceanic crust has the structure it does. (You may care to treat these questions like a series of ITQs, jotting down your answers before reading the text beneath each question. If you're revising this Section, you may find that this will test your understanding and recall of the material.)

❑ What might the significance of the dyke complex be in terms of the formation of the pillow lavas?

■ The dykes are vertical fissures filled with basaltic material, identical in composition to the pillow basalts (at least initially, before the basalts came into contact with seawater). As the dykes underlie the pillow basalts, they could be feeders for the sub-sea basalt eruptions. The fact that they occur in sheets of dykes emplaced into dykes, one after the other, must mean that they form part of either a continuous or an often-repeated process. Magma rising through fissures at ocean ridges erupts at the surface and spreads out laterally. Magma that doesn't reach the surface would cool and crystallize in the fissures, forming solid dykes.

❏ Where did the magma come from that was moving upwards through the dyke fissures and erupting at the surface?

■ There must be a clue from the fact that the thickest part of the oceanic crust sequence — a coarse-grained, layered gabbro — underlies the sheeted dykes. We should expect the gabbro to be linked in some way to the supply of magma.

❏ What might the significance of the layered gabbro complex be in terms of eruption of the pillow lavas and emplacement of the sheeted dykes?

■ The layered gabbro complex is effectively identical in composition both to the dykes and to the basalts prior to their eruption. The reason it is gabbro rather than basalt (coarse crystals rather than fine-grained crystals) is because it cooled much slower than either the dykes or the basalts. As it underlies the dyke complex, it might plausibly represent the chamber which contained basaltic magma waiting to erupt through the fissures onto the sea-floor. Seismic data tell us that this chamber lies at depths of 2–7 km below the sea-floor, so it would have been thermally insulated by surrounding warm oceanic crust. Under such circumstances, we can predict that it would cool slowly and solidify when eruptions cease.

❏ What model can we propose for the eruptive processes that formed oceanic crust?

■ Molten magma must be derived somehow from below the base of oceanic crust. This magma would be stored initially in magma chambers some kilometres beneath the sea-bed, and vented intermittently to the surface via fissures to form the pillow basalts. When a period of eruption ceased, the basaltic magma in the fissures chilled to form dykes. However, the greater volume of this magma would stay in the magma chamber to form gabbro. The nature of the sheeted dyke complex tells us that the process was repetitive, and the magma chamber was an open system. The difference in the seismic velocities occurs mainly because the magma chamber is more deeply buried than the pillow lavas, so it transmits sound waves faster.

❏ What would be the effects of such a magma chamber and its attendant feeder pipes and eruptions on heat flow readings?

■ The heat flow over such an area should be anomalously high, similar to values seen near and over hot-spot volcanoes. We saw in Section 2.2 that heat flow is anomalously high both over hot spots and also over mid-ocean ridges.

❏ What would be the effect of high heat flow and hot rocks on the gravity profile?

■ Hot basalts and gabbros, and particularly molten magma, would be less dense than cool equivalents and nicely explain the observed Bouguer gravity anomaly.

❏ Can we explain the pattern of shallow earthquakes over mid-ocean ridges by our eruption model?

■ We know from plate movement studies that oceanic crust is moving apart at mid-ocean ridges. Shallow earthquakes could be triggered as hot, newly formed, weak crust is pulled apart. The earthquake centres would be shallow because the new oceanic crust is thin. If the process were to be happening almost continuously, there would not be sufficient time for the strain energy which causes earthquakes to build up enough to cause a major earthquake, so low-magnitude earthquakes would be more common. As the crust is pulled apart, it must fracture from its base to its top. Fractures like this could form pathways to allow magma to reach the surface from a reservoir below at the base of the oceanic crust.

❑ Which topographic features on the present-day ocean floor are most likely to be related to the processes of magma generation and venting?

■ The only candidates are the mid-ocean ridges, and aseismic ridges linked to hot spots.

❑ Could oceanic crust be forming at present-day hot spots?

■ Unlikely, as the hot-spot ridges are not markedly tectonically active. They are also associated with large volcanic structures measuring several kilometres vertically, which is not the characteristic structure of oceanic crust. In addition, we see hot-spot chains in places like Hawaii forming structures of limited area extent on top of oceanic crust.

❑ Could oceanic crust be forming at aseismic ridges?

■ Unlikely, as the aseismic ridges are not tectonically active. The process of spreading would reasonably have a seismic signature. They also show no signs of present-day volcanism.

We have now reached a critical point in our deductions. We have eliminated hot spots and aseismic ridges as sources of oceanic crust, and we can explain all the observed geological and geophysical features we see at mid-ocean ridges by a highly plausible model for the formation of new oceanic crust. We can conclude, hopefully beyond reasonable doubt, that oceanic crust must be created at spreading ridges. So far in this Block, our studies have separated the geological and geophysical features of mid-ocean ridges from the known composition of oceanic crust, but here is the link between the two. Oceanic crust has the structure it does because it is produced at ocean ridges by the processes we have just described. The whole system — high heat flow, shallow earthquakes, elevated sub-marine topography and frequently erupting magma chambers — links together to identify the site of creation of new plates. No wonder mid-ocean ridges are called constructive plate margins, for these ridges are the building sites for the oceanic plates themselves!

The evolution of the petrological model is summarized in Figure 2.19. At an ocean ridge, plates are moving apart. A zone of existing oceanic crust is stretched and comes under tension. In turn, this reduces the confining pressure in the mantle and allows melting to take place at depth. Magma rises to occupy the low stress site. This magma is contained several kilometres beneath the ocean ridge. From time to time the magma chamber erupts through vertical fissures, creating new oceanic crust with a characteristic layered structure. If stretching continues, the newly formed lithosphere moves away from the ridge site and the cycle starts again.

❑ But why should there be a 'layered' layer at the base of oceanic crust?

■ In the magma chamber, crystals that nucleate first are dense minerals that sink through the magma. Thus, layers of crystals of almost identical composition form at the base of the magma chamber.

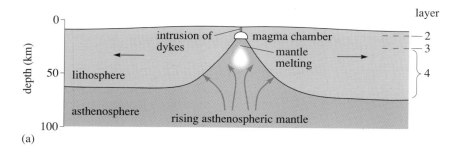

(a)

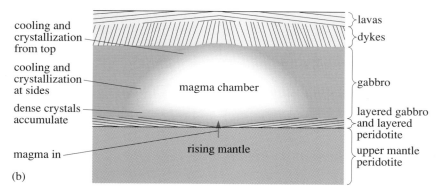

(b)

Figure 2.19 (a) Cross-section showing igneous processes at active ridge crests. Rising asthenosphere fills the gap between separating plates. Some of it melts, ultimately giving rise to lava eruptions at the sea-floor, and forms new oceanic crust. The remainder accretes to the edge of the mantle lithosphere. (b) Diagrammatic enlargement of the axial region of (a). Note that most of the magma crystallizes in the magma chamber itself.

You may recall that we defined the Moho as a discontinuity in seismic velocities, which lies between the gabbros and layered peridotites of our petrological model (Figure 2.19). Where these rocks are found at outcrop, there is unlayered peridotite beneath these gabbros and layered peridotites. Unlayered peridotite is thought to be true mantle material whereas the layered peridotites are strictly part of the crust. It follows that in the petrological model the base of the crust lies between the layered and unlayered peridotites. This is known as the **petrological Moho**, and it is distinctly different from the **seismic Moho** because there is no obvious change in P-wave velocity across it. The seismic Moho lies within oceanic crust, as it is defined petrologically.

Spreading centres and ridge segments

In the preceding Section, we were able to construct a good model for the process at constructive plate boundaries. The critical features beneath the ridges were recharged magma chambers created by tension across the ridge. We can tell from seismic reflection studies across the present-day East Pacific Rise that these chambers are normally 4–6 km wide. There is some evidence that they might be restricted to 10 km wide as a maximum, and that probably they are wider at faster spreading axes than at slower ones. This gives us a good two-dimensional picture — a cross-section through a spreading ridge. But what does the ridge look like in the third dimension? Does the model we have just postulated apply along the entire length of the present-day ridge system?

We can obtain a remotely sensed image of the sites of lava eruption using *GLORIA*-type sonar. This machine, developed in the 1980s, scans the sea-floor obliquely with sonar beams that 'illuminate' the ridge system with sound (Figure 2.20).

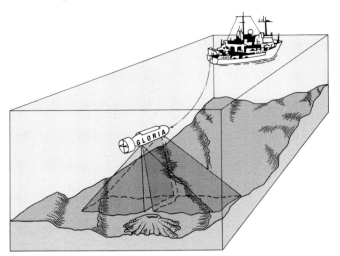

Figure 2.20   *GLORIA* (*G*eological *LO*ng *R*ange *I*nclined *A*sdic) is towed underwater, at a depth of 50 m. In water of 5 000 m depth, *GLORIA* uses a sound beam to scan a zone of the sea-floor about 60 km wide. By steaming on parallel courses, scientists on board the towing ship can obtain overlapping sound images to help interpretation. Sonographs produced by *GLORIA* reveal not only local topography but also submarine canyons, submarine slides and small submarine volcanoes.

Detailed surveys over modern ridges by *GLORIA* and similar systems tell us that individual eruptive centres are no more than about 2–3 km long and that they are often separated from each other along the ridge axis by an inactive gap of about 1 km. Constructive margins are not therefore uniform along their entire lengths. The ridges are segmented at least as far as the eruptive centres are concerned.

Does this mean that the feeder magma chambers are also isolated, and similarly spaced along the axis? It seems not, at least on a scale of kilometres. We are able to image the top of present-day magma chambers using seismic reflection techniques, and we find that seismic reflections from the top of layer 3 are virtually continuous.

In that case, does a single magma chamber run along the entire length of a ridge? Are there any breaks in the magma chambers? Are there any variations in the shape and activity of the ridge along its length which will give us an idea of how continuous magma chambers are at depth? These questions are best answered by considering one well-studied segment of a spreading ridge in detail.

Case Study: The East Pacific Rise between 10°N and 16°N

Detailed study of the East Pacific Rise between 10° N and 16° N reveals two important features, giving more information about the processes that underlie ridges.

ITQ 20

Look at Plate 2.2. This is an image of part of the Pacific Ocean floor in the region of 11° 50′ N 103° 40′ W. Are there variations in shape and form of the East Pacific Rise in this area?

First, ridges are not continuous linear features on a small scale but have small kinks, which on detailed examination prove to be offsets of the ridge by between 2 and 15 km. These kinks or offsets occur at regular intervals along the ridge. On the East Pacific Rise, the straight segments between the kinks are 20–100 km long. In detail, the offsets occur where the axial crest region gradually dies away and a new ridge segment begins to build up alongside it (Figure 2.21). These places where the axial ridge overlaps in this manner have been called **overlapping spreading centres**.

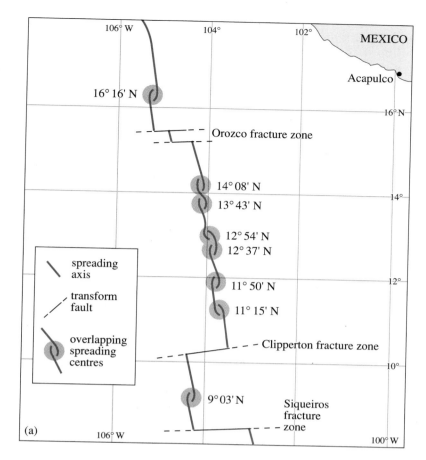

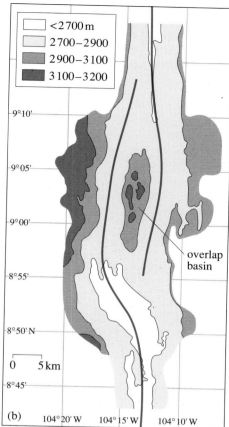

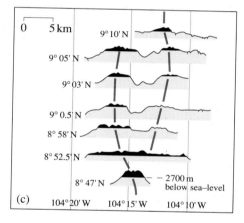

Figure 2.21 (a) Overlapping spreading centres recognized along a 1 000 km length of the East Pacific Rise. (b) Depth contours of the overlapping spreading centre located in (a) at 9° 03′ N on the East Pacific Rise. The axial crests are marked by red lines. (c) Topographic cross-sections through the same spreading centre. The most shallow areas (shown in black) are mainly young lava flows. Vertical scale exaggerated ×4.

Observations from submersibles reveal that young pillow lavas occur on both ridge crests, so it is probable that these crests lie directly over a zone of dyke development. Faults and fractures increase in numbers along each ridge segment towards the overlapping spreading centre. A prominent **overlap basin** is commonly developed between overlapping ridges. One is particularly well developed at 11° 50′ N on the East Pacific Rise and is shown in Plate 2.2.

It is not easy to work out what exactly is going on beneath overlapping spreading centres. The actual processes themselves are concealed from view and too deep to sample physically. We can image them geophysically, but seismic reflection studies in particular are hampered by the presence of molten rock or mushes of partially solidified magma which have peculiar seismic characteristics.

In circumstances such as these, when present-day examples are difficult to study because of the limitations of our current experimental technique, geoscientists often progress by constructing a physical model. This is quite different from the model we deduced from data in the preceding Section. The physical model is an analogue, necessarily scaled down in size. Usually, materials are chosen to represent the scaled-down physical conditions as faithfully as possible.

Overlapping spreading centres have been modelled by observing the behaviour of slits in a solid wax film floating on molten wax (Figure 2.22a). The solid wax sheet is pulled apart at right angles to the slits, which causes them to propagate (Figure 2.22b). As they overlap, the stress field around the ends of the slits changes, and the slit tips curve towards each other (Figure 2.22c). Eventually, one propagating tip links with the other slit (Figure 2.22d) and a continuous slit (ridge) is again developed (Figure 2.22e).

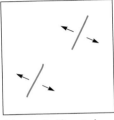 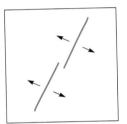   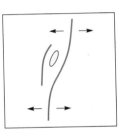

(a) two knife cuts in frozen film; spreading initialized

(b) propagation of spreading centres along strike

(c) spreading centres overlap and curve toward each other

(d) progressive shear and rotational deformation continues until one OSC links with the other

(e) continuous spreading centre is established

Figure 2.22   Wax analogue model of the development of overlapping spreading centres (OSCs).

This model leads us to suggest that continuous ridge segments are underlain by a discrete magma chamber. The chamber is many times bigger than the distance between eruptive points, so several eruptions of pillow basalt are taking place from any one magma chamber. The chambers themselves are not continuous along the ridge, though, and are fed regularly by pulses of molten material from below the plate. When magma pulses in, the continuous segment of ridge propagates. Eventually, these propagating ridge tips will meet, either head-on or (more likely) offset by a small distance. The ridge tips will interact, forming overlapping spreading centres.

In detail, there is a complex array of possibilities; direct hits between ridge segments, ridges missing each other but eventually linking, and ridges missing each other and never actually linking. Features representing each of these have been recognized on fast-spreading ridges throughout the world system. Although the details are not important and you need not remember them, you may care to compare the features predicted in Figure 2.23 with those imaged over the East Pacific Rise, shown in Plate 2.2.

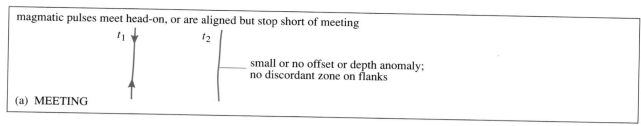

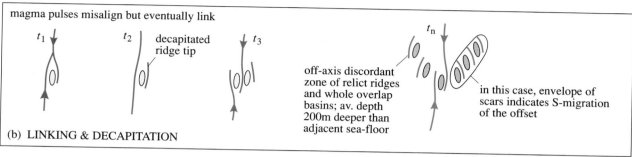

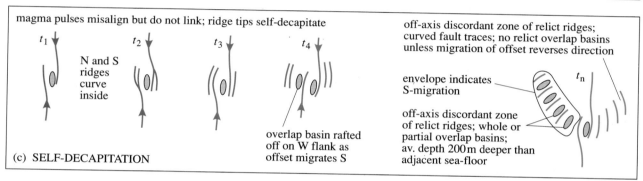

2.3.2 TRANSFORM FAULTS AND FRACTURE ZONES

In the preceding Section, we mentioned that there were two important features which give information about the processes that underlie ridges. Overlapping spreading centres are obviously one, but what is the other?

The ridge system can be offset on a scale altogether bigger than the ridge segments we have so far considered. This is clearly shown on the Smithsonian Map, on each and every one of the currently active ridges. Look at, for example, the East Pacific Rise for 10° N and 10° S of the Equator, where the Pacific Plate abuts the Cocos Plate (north) and the Nazca Plate (south). The north–south-trending red line, which represents the spreading axis, is offset by about 300 km south of the Equator. These features are too large to be overlapping spreading centres and they are also the sites of many shallow earthquakes, arranged in a linear zone between the ends of the spreading ridges. Large-scale displacements of spreading ridges like these are called **transform faults**.

Transform faults are a type of structure that occurs where the ridge is substantially offset, putting oceanic crust of different ages next to each other, and at the ends of ridge segments. They often have offsets of 50 km or more, and partition the ridge system into tectonic units that can persist for millions of years. For example, the Clipperton Transform offsets the East Pacific Rise (at 10° N) by 85 km and has existed for at least the past 9 Ma (Plate 2.3).

What happens to magma chambers beneath the ridge system as it meets a transform fault? Transforms clearly mark changes in the sub-ridge magma chamber. South of the Clipperton Transform (Plate 2.3), the ridge is shallow and wide, and detailed seismic reflection studies show that there is a shallow magma chamber with a top lying about 1.5 km beneath the ridge. In contrast, north of the Clipperton Transform the ridge segment

Figure 2.23 Three possible situations developed where propagating ridge segments meet. The arrows refer to the direction of propagation of the magma pulses, and the labels t_1, t_2, t_3, etc. refer to a time sequence. You don't need to remember the details of this diagram. (S-migration = southward migration.)

is deeper and narrower at the fault. Seismic studies show that the magma chamber is absent for 70 km north of the transform. The composition of erupted basalts is different north and south of the transform, too.

Because almost every detailed structure of the ridge is different on either side of this transform, it seems likely that these two segments are evolving independently. The segment to the south is currently active and swollen with magma while the segment to the north is currently inactive and starved of magma. Transforms must therefore represent major breaks in both the ridge system and ridge magma-supply system. They form a fundamental part of the plate tectonic structural array.

Transform faults have a peculiar type of geometry and are unlike many other faults. Discover the difference for yourself by trying ITQ 21.

ITQ 21

Use Figure 2.24 to answer these questions. It shows two identical geometrical situations in plan view. First, assume that the horizontal thick black line represents a fault and the red line represents some geological feature that has been displaced by movement along the fault. Using the diagram on the left (Figure 2.24a), draw arrows on the fault to show the sense of displacement. Has the feature been displaced leftwards (sinistrally) or rightwards (dextrally)? Now turn the page upside down. Is the sense of displacement the same or different? Next, assume that the horizontal line is a transform fault and the geological feature is a spreading ridge. Using the diagram on the right, put arrows on both sides of the two ridges to show the plate movement directions away from the ridge. Armed with this information, now draw arrows on the fault to show the way the two plates are moving past each other. Are they moving rightwards past each other (dextrally) or leftwards (sinistrally)? Now turn the page upside down. Is the sense of displacement the same or different? Is the sense of movement on a transform fault the same as the sense of movement of a fault that displaces some other geological feature?

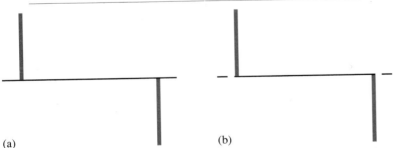

(a)   (b)

Figure 2.24   For use with ITQ 21.

Before the existence of transform faults was known, offsets of the mid-ocean ridges were interpreted by geoscientists as later faults which offset the oceanic ridge system. Parallel sets of faults showing lateral displacement were common on the continents, so they saw no reason why the oceanic equivalents should be any different. The time sequence seemed to be clear; the ridge developed, and at a later stage it was offset by some tectonic process which generated the fractures.

One of the most penetrating of the early plate tectonic insights was made in 1965, when J. T. Wilson recognized the link between the offsetting faults and the spreading ridge. He saw that they were not in fact later faults offsetting the ridge, but a new class of fault that was genetically part of the spreading ridge. These faults grew as part of the plate movement process, and they were the same age as the ridges, not younger than them. Far from being late offsets, they were in fact contemporaneous structures.

To distinguish these new faults, and separate them from other types of faults which also have a sideways displacement (which were already called transcurrent faults), Wilson coined the term *transform fault* ('trans' relates to the lateral translation of plates on either side). The name also related to the fact that lateral movement on the fault is 'transformed' into plate movement at right angles to spreading ridges.

There is another important difference between transcurrent and transform faults, which relates to the ends of the structure. Look again at the ends of the faults in the answer to ITQ 21 (Figure 2.61). As the transcurrent fault in the left-hand diagram displaces a pre-existing feature, it must extend beyond the structure, at least for some distance. With the transform fault on the right, movement along the fault is caused by and related to plate movement. At the ends of the transform fault, the plates are moving in the same direction on both sides. There is not necessarily any relative movement between plates, and no need for the transform fault to continue beyond the ridges. These extensions of transform faults beyond the ridges are no longer necessarily active seismically (check with the Smithsonian Map, particularly at 55° S 130° W). Zones beyond transform faults like these are called **fracture zones**.

Small transform faults may have displacements of several kilometres; the largest transform faults have displacements of the order of thousands of kilometres. The Clipperton Transform on the East Pacific Rise (Figure 2.21) apparently offsets the ridge by 85 km, while the Mendocino Transform off the northwestern coast of California at 125° W 40° N has an apparent displacement of almost 1 200 km. The Mendocino Transform, together with its near neighbour the Pioneer Transform, accommodate a 1 450 km offset of the Juan de Fuca Ridge.

Smaller transforms can be considered as part of the ridge system, and therefore as an integral part of the constructive plate margin system. Larger transforms, however, form the class of margin known as **conservative plate boundaries**, because lithosphere is neither created nor destroyed along them. Whilst transforms are plate boundaries, fracture zones are not.

Case Study: The Clipperton Transform Fault

Plate 2.3 shows in fine detail the Clipperton Transform Fault on the East Pacific Rise at about 104° W 10° N. This particular fault has a marked valley on the south side of the transform, although the fault itself occupies a narrow cleft through generally higher topography further north. This whole zone is one of sharp positive relief compared with ocean floor away from the ridge–transform zone. Most transform faults are major topographic features on the ocean floor, showing a prominent scarp where the level of the ocean floor drops from one side of the transform to the other. Many also have an obvious valley and ridge, although any or all of these features may be absent (Figure 2.25).

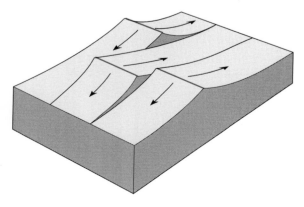

Figure 2.25 The elevation of the ocean floor is different on opposite sides of transform faults and fracture zones, so that escarpments are formed.

ITQ 22

Examine the Clipperton Transform Fault zone shown on Plate 2.3 carefully and attempt the following questions:

(a) Is the transform zone higher or lower than the East Pacific Rise? What is the maximum relief of the zone?

(b) Is there a fracture zone associated with this transform?

(c) Is the sense of movement along the Clipperton Transform dextral or sinistral?

(d) What are the two features labelled RTI?

2.3.3 DESTRUCTIVE PLATE BOUNDARIES

So far, we have considered two of our three principal plate boundary types. Our remaining plate boundary type we have called both convergent and destructive. Are these two types synonymous?

ITQ 23

Look at the Pacific Ocean on the Smithsonian Map and locate as many areas as you can where the Pacific Plate is converging with another plate (use the plate motion arrows to identify plate movement directions). What is the tectonic nature of the plate boundaries in these areas?

ITQ 24

Use Plate 2.1 to identify which areas you identified in ITQ 23 are also associated with 'missing' ocean floor.

The evidence we have gleaned from these two ITQs suggests that convergent plate boundaries are linked with missing ocean floor, and therefore sites of oceanic plate destruction. The case is still far from proven, however. There is a vast body of observation and evidence which helps our case; we should look at some of it now.

You may recall from Block 1 that destructive plate margins are marked by curved ocean trenches, characterized by both deep and shallow earthquakes. These are sites where ocean floor is missing because it has been **subducted** into the underlying mantle. Ocean trenches are the longest depressed linear features on the Earth's surface (the Peru–Chile Trench is 4 500 km long and reaches depths of 7–8 km below sea-level). They are developed on the oceanward side both of active continental margins, like the Andes, and of chains of volcanic islands, called **island arcs**. Volcanism is particularly important on the landward side of trenches; the majority of the volcanoes shown on the Smithsonian Map are situated either in island arcs or on continents immediately landward of an ocean trench.

Destructive margins are mixed zones of both tension and compression. Plates come together and overlap at destructive margins. Destructive margins have a complex of alternative names, including **subduction zones** and **trench–arc complexes**. Plates are being subducted at speeds varying from a slow 20 mm yr^{-1} in the Caribbean to a rapid 120 mm yr^{-1} under the Andes.

We shall start our study of destructive margins by looking at the topographic, geophysical and geological data that characterize subduction zones.

Subduction zones

Here are some observations:

- Some plate margins are characterized by very deep ocean trenches, up to 8 000 m deep, which form the deepest part of the oceans. Plate 2.4 shows these features clearly. The sea-bed slopes relatively steeply into the trenches from both the landward side and the oceanward side. The trenches are continuous for many hundreds of kilometres, occurring both off continents (such as along the west coast of South America) and within oceans (such as the east sides of the Philippines and Tonga).

- Wherever trenches occur, there also occurs a belt of both deep-centred and shallow-centred earthquakes. Shallow earthquakes lie on the oceanward side of the deep earthquakes. In seismic terms, these are the most active tectonic belts on Earth.

- Most subduction zones are associated with chains of active volcanoes. As these chains are typically curved arrays of island volcanoes (such as the Aleutians between Alaska and northeastern Russia), they have been called **island arcs**.

- Whilst the volcanic arcs are areas of high heat flow, zones of particularly low heat flow lie immediately oceanward of them, coinciding with the trench.

We shall now investigate what processes are taking place in subduction zones by generating a 'thought model', just as we did for ocean ridges. This requires us to think through all the evidence, drawing on more data as required, to come up with a model for subduction processes. Indeed, it would be better to come up with several models if possible, and test each against all the available data. If we eliminate all impossible models, whatever remains must be close to the truth. We shall use the question and answer technique as in the previous Section.

❑   Why are the earthquake centres arranged so systematically in subduction zones?

■   To answer this, we must first remember how earthquakes are caused. Because we can detect the time of arrival of pressure waves generated by the earthquake at various listening stations around the globe, we know where an earthquake is focused. This is the site of sudden energy release. When we locate earthquake centres on land, they are often associated with known rock faults. (One famous example is the San Andreas Fault zone of California, which is the most seismically active area of the USA.) Earthquakes, then, are associated with movements of rock bodies past one another.

The data set presented on the Smithsonian Map suggests that there are two belts of earthquakes; shallow earthquakes lie oceanward of deep landward earthquakes. This is really only a feature of the way that this particular data set is presented. On the Smithsonian Map, earthquake foci are separated into those above and those below 60 km deep. The reality is that the foci of earthquakes become progressively deeper away from the trench; there is no real 'jump' from above to below 60 km.

Look at Figure 2.26 which shows the earthquake foci around the Tonga Trench. There are very few earthquakes oceanwards of the trench bottom. Shallow earthquakes, with foci at 100 km deep or less, lie mostly within 100 km west of the trench. They give way westwards to progressively deeper earthquakes, and the deepest earthquakes lie over 600 km to the west of the trench. These earthquake foci, at 600 km deep, must be well down into the mantle.

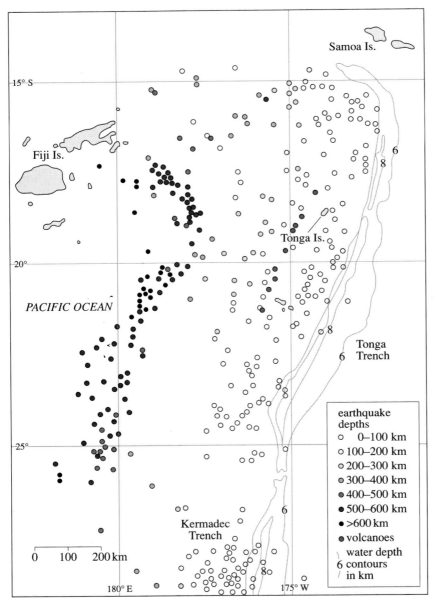

Figure 2.26   Earthquake foci associated with the Tonga Trench. Earthquake centres (foci) get progressively deeper away from the trench. They must be associated with a broad plane of rock fracture dipping westwards.

The Tonga earthquake data tell us that there is a plane of rock fracture lying between Tonga and Fiji, which generates earthquakes. These earthquakes happen continually, suggesting that the rock is continually failing along this plane. This in turn suggests that the two rock bodies on either side of the plane move continually (or frequently) past each other. From the distance and depth data presented on Figure 2.26, we can work out the dip of the failure plane.

ITQ 25

What is the dip of the failure plane west of the Tongan Trench?

ITQ 26

Is this angle constant along the whole length of the Tongan Trench?

The Tongan Trench has a pattern of earthquake foci typical of convergent margins across the Pacific. Figure 2.27 shows the distribution of earthquakes beneath Japan. The dip of the Japanese failure plane is about 1 in 3, or almost 20° to the west-northwest. Other trenches show similar patterns.

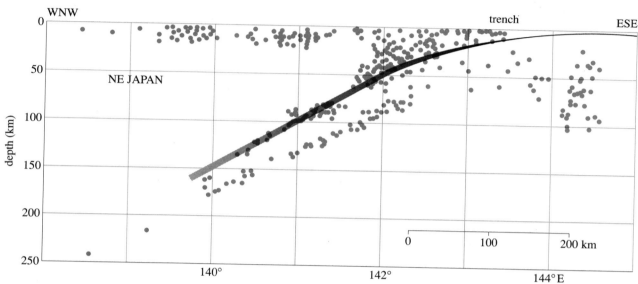

Figure 2.27 Distribution of earthquakes beneath NE Japan.

Compare the dip directions of these two failure planes (Tonga and Japan) with the plate movement vectors shown on the Smithsonian Map. At Tonga, the Pacific Plate is moving towards the Indo-Australian Plate in a west-northwest direction (but note that the Indo-Australian Plate is also moving northwards at the same time). At Japan, the Pacific Plate is moving northwestwards towards the Eurasian Plate. To within about 20°, the plate movement direction is parallel to the dip direction of the seismically active fault.

There is a failure plane triggering active earthquakes associated with every trench. These seismically active zones are called **Wadati–Benioff zones**, after the discoverers. Earthquakes on them typically extend from the surface at the trench to around 680 km depth, where they stop. All Wadati–Benioff zones show frequent seismic activity, which itself shows that the failure plane that causes them is itself continually active.

❑ Does the depth of the ocean floor relate to this failure plane?

■ We know already that the ocean floor drops suddenly and dramatically into oceanic trenches on the ocean side of Wadati–Benioff zones. Detailed study of depth data near trenches shows that there is also a small but significant rise or bulge about 150 km oceanwards of trenches. In Figure 2.28, 35 topographic cross-sections through different Pacific trench systems have been stacked together to show the typical depth variation around trenches. Water depth on the ocean side of the trench is remarkably regular, varying between 5 km and 7 km over the deep ocean. The exact depth of the ocean floor seaward of the trench is governed by the age of the oceanic crust. The deepest parts of the trenches are consistently 2 km deeper than their surrounding oceans, yet always there is a bulge of about 0.5 km immediately oceanward of the trench.

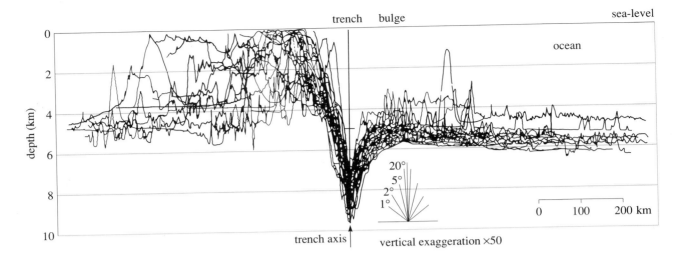

Figure 2.28 Topographic cross-sections through 35 Pacific subduction zones, stacked one upon the other. Whilst water depths are very variable landward of the trench, the water depth is quite uniform on the oceanward side (depending on the age of the oceanic crust). There is a clear shallowing or bulge, centred about 150 km oceanward of the trenches.

The presence of the bulge gives us some important information about the oceanward side of the trench and the nature of the failure plane. The bulge is caused by flexure of comparatively rigid oceanic lithosphere at the subduction zone. The details are beyond the scope of this Block, but you can get a good idea of the principle by trying this simple experiment. Place an ordinary plastic 30 cm ruler on the edge of a bench or table top so that about one-third of it protrudes over the edge (Figure 2.29). Hold down the table-top end of the ruler firmly, but make sure you only hold the extreme tip of the ruler. Now press down gently on the free end. The ruler will flex, and the part that was flush with the table top will rise slightly off it. As if by magic, you have created a bulge like those near ocean trenches!

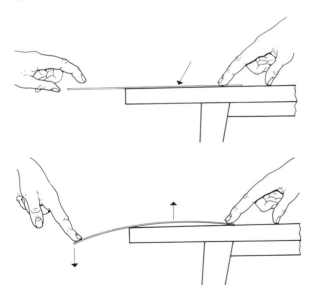

Figure 2.29 Flexing a ruler over the edge of a table. A bulge is created because of the rigidity of the ruler, rather than any effect of the table edge or the amount of downward pressure.

The important part of this experiment is how rigid the ruler is. Try the same experiment with a thick steel bar and you will not get any bending; try it with Plasticine or warm toffee and you will create a bend but no bulge. This is because the ruler is of just the correct rigidity to flex; the steel bar is too rigid and the toffee is too soft. The existence of this bulge helps to show that lithospheric plates are rigid.

At trenches, oceanic lithosphere is being deformed and flexed downwards. The slope on the ocean side of the trench is caused by the flexing of the lithosphere. The bulge is explained in terms of the rigidity of the oceanic plate, but it also suggests that part of the rigid oceanic plate must extend beneath the landward plate, just as part of the ruler protruded beyond the table top.

❑   If part of the oceanic plate extends underneath the landward plate, what causes the earthquakes?

■   There is a simple way of combining the presence of a slab of oceanic crust under the landward plate and the existence of a zone of failure dipping landward down to about 700 km. It is possible that the top of the descending oceanic crust is moving against the lower part of the lithosphere under the landward plate. This suggests that a whole slab of oceanic lithosphere is being driven down into the mantle beneath the overriding landward plate, creating ever-deeper earthquakes by relative movement between the two plates.

❑   Does the rest of our geophysical data set support the existence of a downgoing slab of oceanic material?

■   There are many additional lines of geophysical evidence which support this model. First, there is a difference in type between the earthquakes generated near the trench and those generated deeper down. The shallow earthquakes almost all originate from stretching or tensile stresses, as you might expect if the outer arc of the downgoing slab were stretched during flexure. (You can see this well by bending a Mars bar — try it!) The deeper earthquakes originate from compressional stresses, again as might be expected if resistance between plates was trying to stop the descent of the oceanic slab. Detailed research has shown that the deepest earthquakes are not generated by slab movement but by internal deformation, and the majority of deep earthquake foci don't actually lie at the top surface of the slab but about 30–40 km beneath the top.

The bulge oceanward of the trench is marked by a positive gravity anomaly while the trench itself is marked by a large negative gravity anomaly. Figure 2.30 shows the free-air gravity profile across the Aleutian trench–arc system in the North Pacific, which is typical of Pacific Plate subduction zones.

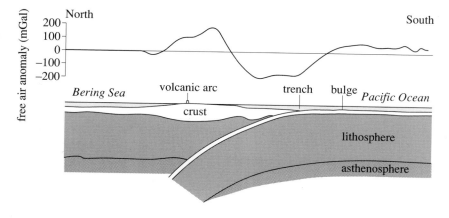

Figure 2.30   Free-air gravity anomaly over the Aleutian trench–arc system, related to a schematic section through the lithosphere assuming the Wadati–Benioff zone is created by a descending lithospheric slab. Note the positive gravity anomalies over the bulge and arc, and the negative anomaly over the trench itself.

The positive anomaly over the bulge and the much bigger one over the volcanic arc landward of the trench are caused by dense oceanic and volcanic rocks. The positive anomaly over the volcanic arc itself means it cannot be in isostatic equilibrium, and must be being supported somehow. By contrast, the anomaly over the trench, also underlain by oceanic crust, is too low for the trench to be in isostatic equilibrium. It must be being held down by some force. The negative anomaly represents a mass deficit which can probably be explained by low-density rocks within the oceanic crust. This isostatic imbalance, with the arc held up and the trench held down, is due in part to the flexural rigidity of the downgoing slab.

Thirdly, the heat flow variations across the trench–arc system may be explained by considering the thermal effect of a descending cold slab. We saw in Block 1 that trenches are sites of low heat flow compared to normal oceanic crust. If a thick slab of cold oceanic crust of low thermal conductivity is subducted into hot mantle, it will take an enormously long time to heat up and reach thermal equilibrium with hot mantle. Relatively easy calculations tell us that an oceanic slab 125 km thick (crust plus lithospheric mantle) will take several hundred million years to reach equilibrium with mantle rocks at 500 km depth. We know, too, how long it would take material from the surface to reach those depths; a slab descending at $50 \, mm \, yr^{-1}$ would take about 14 Ma to reach depths of 500 km. A subducting cold slab will have a major and lasting thermal effect, decreasing the surface heat flow in the surrounding area.

In Block 1, Figure 1.101 shows that the surface heat flow modelled from a cold descending slab closely parallels the observed heat flow in trench regions. The descending slab depresses heat flow in the trench region for at least 100 Ma after subduction commenced. Landward of the trench, heat flow values become restored to 'normal' over a period of about 100 Ma. Figure 1.101 suggests that heat flow equilibrium has been achieved in the Japanese Islands.

ITQ 27

Using the data in Figure 1.101, how long ago do you think the Japanese subduction zone was established?

ITQ 28

From your answer to ITQ 27, and using Plate 2.1, what then is your best guess for the minimum age for the Pacific Plate?

So seismic, gravity and heat flow considerations all support our assertion that convergent margins are also destructive plate boundaries. At these boundaries, oceanic crust is descending steeply into the mantle beneath an adjoining plate. Oceanic crust, formed at divergent margins millions of years ago by transfer of material from the underlying asthenosphere to the overlying lithosphere, is being returned by subduction back into the asthenosphere at convergent margins.

Case Study: Lesser Antilles subduction zone

Processes at destructive margins are complex, as we have seen. There are many physical features of these regions that are directly associated with plate subduction but that do not, at first sight, appear to be related. To put together all the geological features that typify subduction zones, we shall look in detail at a real example of a particularly well-developed present-day ocean subduction zone, the Lesser Antilles subduction zone.

The Lesser Antilles arc lies in the southeastern Caribbean off the northern coast of South America (Figure 2.31a). You can locate the arcuate zone of earthquakes and volcanoes on the Smithsonian Map at 60° W 15° N.

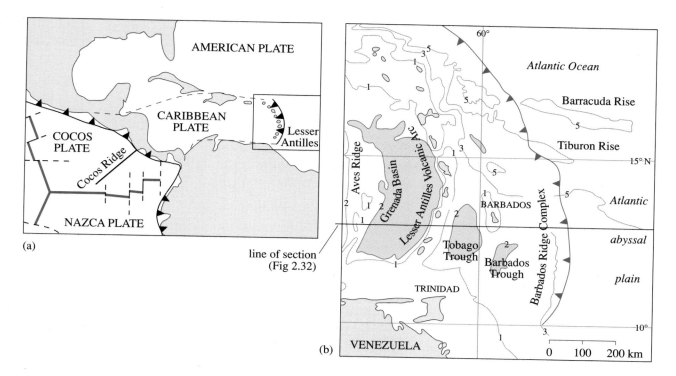

At the Lesser Antilles subduction zone, the Atlantic Plate is descending under the Caribbean Plate at about 20 mm yr^{-1}. We can (just about!) see from the pattern of earthquake foci on the Smithsonian Map that the deeper earthquakes lie westwards of the more shallow ones, so like the Tongan Trench, the subduction zone must be dipping westwards. The actual dip amount is some 28° to the west. The arc is convex eastwards, towards the subducting plate. Movement between the Caribbean Plate and the adjacent North American Plate to the north of the arc is taken up by a major sinistral transform fault.

We will traverse westwards from the Atlantic Plate through the arc system into the Caribbean Plate. Figure 2.32 is a detailed cross-section through the whole subduction zone, and we shall refer to that diagram throughout our case study. In this diagram, there are details of rock type and magma evolution which we have not as yet covered, and will be unclear to you at the moment. Bear with us, for these topics will be discussed fully in Blocks 3 and 4.

Figure 2.31 Location (a) and main structural features (b) of the Lesser Antilles subduction zone. Lines with teeth indicate subduction zones. In (a), red lines indicate spreading ridges. Dashed lines are transform faults and fracture zones.

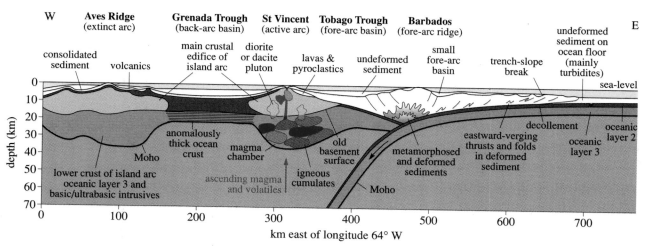

To the east of the subduction zone there is normal oceanic crust, with the typical three-layered structure discussed earlier. In this part of the Atlantic, it is overlain by a considerable amount of sediment, labelled 'mainly turbidites' on Figure 2.32.

Figure 2.32 A detailed cross-section interpreting the geology through the Lesser Antilles subduction zone.

53

ITQ 29

It was noted earlier that sediment cover is usually thin in the deep oceans. Why might this area be different?

The expected trench is also missing. It probably did exist at one stage in the evolution of the subduction system, but it too has now been filled with sediment derived from South America. Its presence today can only be detected by a belt of negative free-air and Bouguer gravity anomalies that lie to the east of Barbados. The deep-water trench has been replaced by a thick ridge of sediment of Tertiary age that has been accumulating on the trench site for more than 50 Ma. This is the Barbados Ridge. The island of Barbados is formed by the top of the sediment pile where it emerges above the ocean. The sediment is over 20 km thick in the thickest part of the Barbados Ridge.

Why is the sediment so thick? Under normal conditions, several kilometres of sediment might accumulate in a major river delta. If this delta were to spread into a trench at a destructive boundary, as is the case here, the sediment might fill the 8 km 'hole' that we would expect to find at the deepest part of the trench itself. As the sediment is denser than the water it replaces, the crust will become loaded at the site of deposition and will sink isostatically. This allows room for more sediment to accumulate, which in turn produces more subsidence. Nevertheless, this process will still only account for about half the observed thickness of sediment. Some other process must be operating where the subducting plate dips beneath the overriding plate.

This additional process at work is **rock deformation** by folding and thrusting, driven by the relative movement between plates. Look again at Figure 2.32. The Barbados Ridge is made entirely of sediment, occupying the trench site. At the east margin of the ridge, this sediment is being transported westwards by the ever-moving Atlantic Plate. On the west side of the ridge, the sediment is lying directly on the (relatively) stationary Caribbean Plate. Even as more sediment gets deposited on the Atlantic Plate, it is driven westwards into the pre-existing sedimentary pile. The act of compressing the sediment achieves two results — it thickens the sedimentary pile and it 'scrapes off' sediment which was deposited on the Atlantic Plate and 'plasters' it onto the bottom of the sedimentary ridge that is effectively sitting on the Caribbean Plate. Sediment is said to be **accreted** onto the overriding plate from the subducting plate, and is effectively being transferred from one plate to the other. The whole sedimentary structure is called an **accretionary prism**, and is a typical feature of the trench area of subduction zones.

Accretionary prisms are composed of slices of deep-water sediment, impacted into each other. The slices are separated by faults that allow thickening by stacking — these are known as **thrust faults**. Often the slices are deformed internally by **folds** which show their derivation by their internal geometry. In Figure 2.32, the folds and thrusts are referred to as 'eastward verging' which shows that they formed by relative eastward movement of the sediment compared to the moving plate. (In reality, it was the plate moving westwards underneath the sediment that created the eastward-verging structures.) If you are sceptical that this could happen, clear snow from a path after the next heavy snowfall in your area, and look in detail at the way the snow stacks in front of the shovel.

Thick piles of sediment like this accretionary prism are necessarily compressed and heated at their bases. Heat and pressure changes the crystalline structure of the minerals that make up the sediment, in a process known as **metamorphism**. Block 4 goes into some considerable detail about metamorphic effects, but for our purposes we can simply note that the base of the Barbados accretionary prism is described on Figure 2.32 as 'metamorphosed and deformed sediments'.

Further west, about 150 km west of Barbados, lies a curved chain of volcanic islands, the Lesser Antilles themselves (clearly shown in Figure 2.31b). This is the **island arc**. Figure 2.32 shows the geological details of this island arc to be extremely complex, but we shall deal only with the principal features of the arc because island arc volcanism will be explored fully in Block 4.

The crust is thicker at island arcs. Figure 2.32 shows that the crust under St Vincent, one of the Lesser Antilles, is over 40 km thick and over six times as thick as the descending oceanic plate. If thin oceanic crust once underlay St Vincent, it has now been supplemented by vast volumes of additional material. The additional material is almost all igneous rock. Quite unlike the Barbados Ridge, which was formed from sediment deposited in the trench, the Lesser Antilles arc is a volcanic complex. Lavas and **pyroclastic rock** (fragmental rocks formed from explosive eruptions) form much of the arc. These lavas are intruded by **plutons**, or large bodies of magma which have cooled slowly at depth. Geoscientists know from geochemical investigations that the plutons are genetically linked to the lavas and pyroclastic material. The vast thickness of the crust here compared with oceanic crust seems to represent the superstructure and infrastructure of a huge volcano.

More strictly, the arc represents a line of volcanoes spaced at a critical position in the subduction complex. Look at Figure 2.31b and at the volcanoes in other island arcs on the Smithsonian Map (particularly the Lesser Antilles, the Tongan area and the Aleutians between Siberia and Alaska). The critical value is not how far the volcanoes are apart, for that seems to vary from arc to arc. Instead, the critical observation is that volcanoes in the arc lie at a constant distance away from the trench, for any given arc.

Why should this be? To answer this question, we need to understand what is happening at depth under the subduction zone. The descending slab leaves the surface at the trench and dips under the upper plate. We know from our studies already, particularly in the Tonga region, that the subducted slab continues to slope steadily downwards under the upper plate for over 500 km from the trench. The Lesser Antilles slab is dipping at about 28° (Figure 2.32 is schematic in this respect). When it is 150 km from the trench, underneath St Vincent, its upper surface will be about 80 km beneath the surface (150 km $\times \tan 28° = 79.8$ km). If this calculation is repeated for many island arcs, with slabs descending at different angles, we find that the volcanoes always form above a point when the descending slab underneath reaches a depth of around 100 km. The volcanoes that form island arcs are a result of magma generation at depth, as we shall discover in Block 4.

Figure 2.32 shows that the Lesser Antilles arc must have migrated eastwards with time, because to the west of the currently active island arc we find an extinct island arc called the Aves Ridge. The structure and composition of the Aves Ridge is similar to the Lesser Antilles arc, the only difference being that the Aves Ridge is no longer currently active. Rock samples dredged and cored from it are typically between 60 and 70 Ma old. We can tell therefore that the Atlantic Plate must have been at about 100 km beneath the Aves Ridge about 60 Ma ago, although it is clearly deeper than that under the Aves Ridge now. We cannot tell from this fact alone whether the trench–slab system has migrated towards the east, or whether the subduction zone has become steeper with time. Either would cause the migration of the arc to the east. In fact, detailed investigations of the Barbados accretionary prism show that the trench has migrated eastwards significantly over the past 50 Ma.

There is one final element of the Lesser Antilles subduction zone to consider. Between the Aves Ridge and the currently active arc lies the Grenada Trough, labelled on Figure 2.32 as a **back-arc basin**. This is a small basin bounded to the east by the currently active island arc and westwards, in this case, by an extinct arc. The Grenada Trough is floored by oceanic crust, identical in composition but, in this case, much thicker than normal oceanic crust. Many oceanic convergent plate margins have back-arc basins (sometimes called **marginal seas**) which are floored by oceanic crust, though generally this crust is 2–3 km thinner than 'normal' oceanic crust.

ITQ 30

Does the existence of oceanic crust in a back-arc basin suggest tension or compression? Would you expect a convergent margin to experience compressive or tensional stresses?

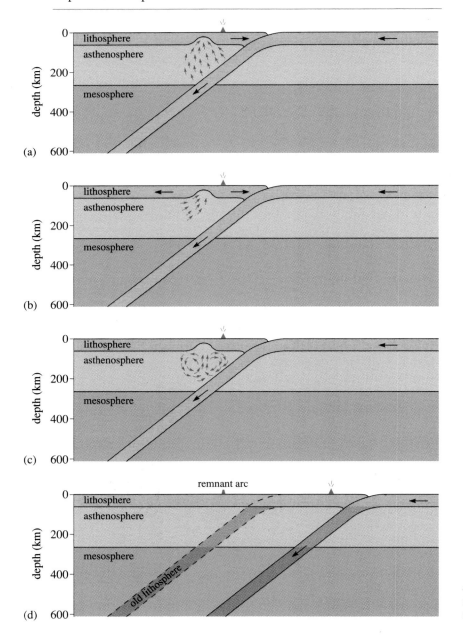

Figure 2.33 Some proposed models for the generation of back-arc basins and marginal seas.

There are two problems concerning the generation of back-arc basins: why should they commonly be floored by generally thin oceanic crust, and why should a zone of tension develop in an otherwise compressional situation? At the time of writing (1992), these problems haven't been solved completely, but several suggestions have been made based on the sort of empirical modelling we ourselves have used (Figure 2.33). It is possible that bodies of mantle-derived basic magma are emplaced forcefully as plutons, causing extension, or it is possible that pre-existing tensional stresses permit the emplacement of plutons (Figure 2.33a, b). In either circumstance, these magma chambers erupt to generate recognizable oceanic crust in much the same way as is proposed for spreading centres.

If a tensional regime permits the emplacement of plutons, why is the zone experiencing extension? Some geoscientists prefer to believe that the downgoing slab sets up subsidiary cells of convecting mantle above the lower parts of the slab (Figure 2.33c), which cause extension by opposed flow. Currently popular models centre around a **trench suction force**, pulling the overriding plate towards the trench and subjecting the back-arc area to extension. This force might be related to an increase of slab dip at depth or to 'roll-back' of the downgoing slab, where the slab retreats from the overriding plate under the pull of gravity (Figure 2.33d). Whatever the reason, once back-arc spreading starts, the trench system becomes decoupled from the landward plate.

In the case of the Lesser Antilles, the roll-back model (Figure 2.33d) is supported because this fits with the suggestion (above) that the trench has migrated eastwards significantly over the past 50 Ma.

2.3.4 COMBINED PLATE BOUNDARIES

So far, we have treated constructive, conservative and destructive margins as distinct and separate plate boundaries. However, the character of a plate boundary depends on its direction of movement relative to the orientation of the boundary. If we look at the Pacific Plate on the Smithsonian Map, we can see that there are many places where the plate motion is not exactly at right angles to the plate boundary, nor exactly parallel to it. In areas like the western end of the Aleutians, for example, movement is oblique to the boundary. Boundaries like these display both the lateral displacement of conservative margins with either subduction or spreading features. Where a plate boundary displays features of more than one type, it is referred to as a **combined plate boundary**.

When movement along transform faults is not exactly parallel to the fault line, these cannot be purely conservative plate margins, but must have a component of constructive or destructive motion. Such behaviour does in fact characterize many transform faults. Those with a constructive or spreading component are called **leaky transforms**. Because leaky transforms have a spreading component, basaltic magma will be erupted along them so that there is a 'leakage' of magma from the upper mantle along the faults.

Just as oceanic plates do not always move exactly parallel to transform faults, so they are not always subducted perpendicular to trenches. This situation must occur, for example, along the changing boundary between the Pacific Plate and the North American Plate. Where a single trench consumes lithosphere from two different plates, it may evolve to form a transform fault (Figure 2.34). This is excellently displayed by the Alpine Fault in New Zealand.

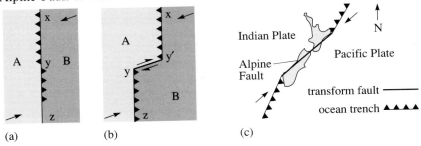

(a)   (b)   (c)

Figure 2.34   The evolution of a transform fault at a destructive margin. The arrows show the relative direction of motion and are on the plates being consumed. In (a), plate A is consumed between y and z and plate B is consumed between x and y. As time passes (b), the trench evolves to form two trenches, xy′ and yz, joined by a transform fault between y and y′. A sketch map of New Zealand (c) shows that the Alpine Fault is a transform fault of this type.

2.3.5 TRIPLE JUNCTIONS

All of the plate boundaries discussed so far are junctions between two plates. However, there are many localities where three plates are in contact, and these are termed **triple junctions**. Triple junctions between three ocean ridges, such as that in the South Atlantic between the African, South American and Antarctic Plates, are known as ridge–ridge–ridge, or RRR triple junctions. Although other types of triple junction are less common, a similar notation can be used to identify triple junctions involving ocean trenches (T) or transform faults (F). A ridge–ridge–trench junction would be termed an RRT triple junction, for example. The ordering of the letters is not significant.

Despite all the geometric possibilities of fitting together three plate margins and their relative motions, there are actually only 10 possible triple junctions. Some of these, such as RRR junctions, are termed **stable junctions**, which means that they maintain their shape through time. However, some can only exist for a moment in time (geologically speaking!) before evolving to another configuration, and these are termed **unstable**. Figure 2.35 shows the evolution of three triple junctions over time.

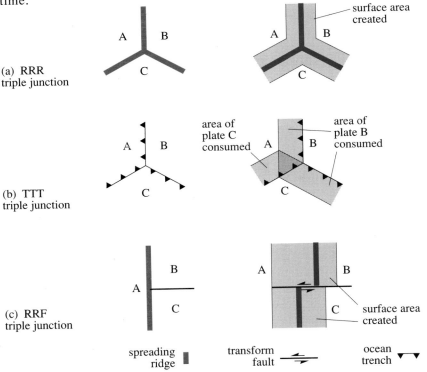

(a) RRR triple junction

(b) TTT triple junction

(c) RRF triple junction

spreading ridge   transform fault   ocean trench

Figure 2.35   The evolution of triple junctions with time. (a) A triple junction involving three ridges (RRR triple junction) is always stable, and the magnetic anomalies within the surface area created have Y-shaped patterns around the spreading ridges. (b) A triple junction between three trenches (TTT) is almost always unstable except in the special circumstance shown here when the relative motion of plates A and C is parallel to the plate boundary between B and C. (c) A triple junction between two ridges and a transform fault (RRF) can exist only for a short instant in geological time and decays immediately to two FFR stable junctions.

The RRR junction shown in Figure 2.35a is always stable, regardless of the relative rates of spreading at each of the three ridges. The TTT junction in Figure 2.35b is basically unstable except if, by coincidence, the movement rates are the same and if the direction of subduction of plate C below plate A is exactly parallel to the boundary between plates B and C. When the relative motion of plate C is not parallel to the boundary between plates B and C, the triple junction is never stable. The triple junction in Figure 2.35c is an RRF junction, and is unstable because there is relative motion between plate B and plate C. The RRF triple junction evolves immediately to form two RFF junctions. FFF and RRF junctions are always unstable.

Only seven types of triple junction exist during the present plate tectonic configuration. Before reading on, you may care to find these out for yourself from the Smithsonian Map! If not, these are RRR (e.g. in the South Atlantic, the Indian Ocean and west of the Galápagos Islands in the Pacific), TTT (Central Japan), TTF (off the coast of Chile), TTR (off Moresby Island, western North America), FFR (on most oceanic ridges), FFT (junction of the San Andreas Fault and the Mendocino Transform off western USA) and RTF (southern end of the Gulf of California).

If RRR triple junctions are the most stable form of triple junction, do RRR triple junctions exist as 'fossils' in old ocean floor?

ITQ 31

What evidence would indicate a former RRR junction in an area of old ocean floor?

It was very exciting when 'bent' magnetic stripes were first discovered in the North Pacific during the late 1960s. Near to the Aleutian Trench, approximately northward-trending magnetic stripes bend sharply towards the northwest by about 70° (Figure 2.36). This bend is called the Great Magnetic Bight of the northeast Pacific, and the magnetic stripes formed between 63 and 75 Ma ago. The simplest interpretation of this feature is that it represents a pre-existing RRR junction.

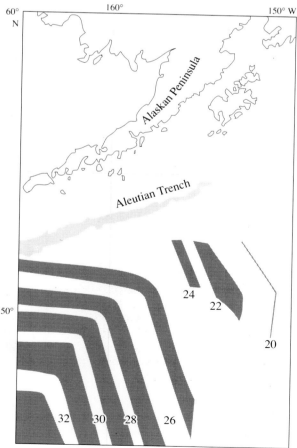

Figure 2.36 Magnetic anomalies in the North Pacific. The lineations are numbered 20–32, and by convention the numbers increase with age. The group numbered 25–32 bend through an angle of 65°–75°. Three unnumbered lineations meet within the Aleutian Trench.

However, only the one limb of the 'Y' set of magnetic anomalies remains, so where is the other part of the RRR junction? The Great Magnetic Bight runs into the Aleutian Trench, so the remaining parts of the triple junction have probably been subducted at the destructive plate margin between the Pacific and American Plates. Such is the fate of the evidence of past plate tectonic activity.

SUMMARY OF SECTION 2.3

- There are three principal plate boundary types: constructive or divergent plate boundaries, destructive or convergent plate boundaries, and conservative plate boundaries.

- Divergent plate boundaries are characterized by a spreading axis, formed of several spreading centres. These occur within oceans for the most part, but rarely are they centrally situated in any ocean (the Mid-Atlantic Ridge is an exception). New oceanic crust is being formed at these centres.

- Full spreading rates at these boundaries vary from a slow rate of $10 \, mm \, yr^{-1}$ to a rapid rate of $90 \, mm \, yr^{-1}$. The form of the ridge, particularly the depth of any median valley, is dependent on the spreading rate.

- Spreading axes are in overall isostatic equilibrium with the surrounding oceanic crust, but contain lower-density material.

- Oceanic crust formed at spreading axes records an eruptive story. Basaltic lavas form new sea-floor. These are fed through sub-vertical fissures which when abandoned contain coarser-grained basaltic rock. A basaltic magma chamber lies underneath the feeder pipes. When abandoned, it solidifies to form gabbro and peridotite.

- The Moho defined from seismic velocity contrasts is at a slightly shallower depth from the Moho defined petrologically.

- Magma chambers do not run along the entire length of a spreading axis, but rather occur at discrete intervals.

- The ridge system is partitioned on a larger scale than spreading centres, by transform faults. These faults offset spreading axes, often by tens or hundreds of kilometres. Transforms are conservative plate boundaries in an oceanic setting.

- Transform faults are distinctive because their sense of movement is opposite to the apparent displacement of the spreading axis. This shows that they do not displace the axis, but form an integral part of the spreading system.

- At convergent plate boundaries, oceanic crust is being destroyed as it is subducted into the mantle.

- Subduction zones are characterized by deep ocean trenches and by zones of earthquake foci related to the position of a slab of subducting oceanic crust. The dipping seismic zones are called Wadati–Benioff zones.

- Bulges in the plate oceanward of trenches and the trenches themselves tell us that the oceanic plate behaves rigidly as it dips under the overriding plate.

- Heat flow recorded on the Earth's surface around subduction zones is consistent with that modelled by a descending cold oceanic slab.

- A typical oceanic subduction zone consists of (from oceanwards to landwards) a slight bulge in the oceanic plate, a deep trench which may in certain circumstances be entirely filled with sediment, an accretionary prism, a volcanic island arc and a back-arc basin.

- Conservative plate boundaries may show a component of spreading (leaky transforms) and volcanic activity, or they may show a component of plate subduction.

- Where plate boundaries meet, triple junctions are formed. At these sites, three plates are in contact with one another. Most such junctions are unstable and disappear with time. Ridge–ridge–ridge (RRR) junctions are characteristically stable and exist both at the present day and as 'fossils' in older parts of oceanic plates.

OBJECTIVES FOR SECTION 2.3

When you have completed this Section, you should be able to:

2.1 Recognize and use definitions and applications of each of the terms printed in bold.

2.8 Describe the geological and geophysical parameters that identify and distinguish constructive, destructive and conservative plate boundaries.

2.9 Identify the tectonic processes that are happening at each of the three main types of plate boundary.

2.10 Understand what happens when plate boundaries interact and predict the outcome of evolving plate boundaries.

Apart from Objective 2.1, to which they all relate, the 16 ITQs in this Section test the Objectives as follows: ITQs 16, 17, 22, 24, 27 and 28, Objective 2.8; ITQs 20, 21, 23, 25, 26 and 30, Objectives 2.8 and 2.9; ITQ 31, Objective 2.10.

You should now do the following SAQs, which test other aspects of the Objectives.

SAQS FOR SECTION 2.3

SAQ 4 (*Objectives 2.1, 2.2 and 2.8*)

How could geophysical data from earthquakes help geoscientists to distinguish between different types of plate boundary?

SAQ 5 (*Objectives 2.1, 2.8 and 2.9*)

By ticking the appropriate spaces in Table 2.2, indicate which of the listed geological, geophysical and geochemical features occur at each different ocean plate boundary type. (Note that some features occur at more than one boundary type.)

Table 2.2   For use with SAQ 5.

Geoscientific feature	Destructive margin	Constructive margin	Conservative margin
Weaker earthquakes (mag. < 7.5)	✓	✓	✓
Stronger earthquakes (mag. > 7.5)	✓	✗	✗
Shallow-focus earthquakes (< 60 km)	✓	✓	✓
Deep-focus earthquakes (> 60 km)	✓	✗	✗
Active volcanoes	✓	✓	✗
Wadati–Benioff zones	✓	✗	✗
Transform faults	✗	✗	✓
Offset magnetic stripes	✗	✗	✓
Missing magnetic stripes	✓	✗	✗
Higher heat flow	✗	✓	✗
Lower heat flow	✓	✗	✗
Median rifts	✗	✓	✗
Magma generation	✓	✓	✗
Accretionary prisms	✓	✗	✗
Fracture zones	✗	✓	✓
Island arcs	✓	✗	✗

SAQ 6 (*Objectives 2.1 and 2.10*)

Why are some ocean plate triple junctions stable while others are not?

2.4 CHANGING OCEAN PLATE BOUNDARIES WITH TIME

We already know enough about plate tectonics to predict that plate boundaries will change with time. The triple junctions considered in the previous Section obviously do, but so do most plate boundaries because plates and plate boundaries are in constant motion over the Earth's surface. Ocean plate boundaries reconfigure because their triple junctions evolve and because ridges and transforms become subducted. The common theme is the time dimension; sooner or later, all plate boundaries must change. The present-day plate configuration has not existed for ever.

In this brief study of the complex pattern of plate boundary evolution, we will focus on two: the changes most spreading ridges experience with time, and the changes in plate boundary configuration that take place due to subduction.

2.4.1 CHANGES AT SPREADING RIDGES WITH TIME

In many cases, spreading at constructive boundaries appears to be symmetric. The widths of parallel magnetic stripes indicate that the rate of formation of new oceanic plates is the same on either side of an ocean ridge. But there is no particular reason why sea-floor spreading should be symmetrical. Indeed, when many magnetic profiles (like that in Figure 2.10) for both sides of an ocean ridge are examined in detail, evidence of irregularities due to **asymmetric spreading** are found. For example, in the Southern Ocean the Australian and Antarctic Plates are presently moving apart with half-spreading rates of about 30–40 mm yr^{-1} as a result of sea-floor spreading at the Australian–Antarctic Rise. A detailed examination of submarine topography and magnetic patterns has shown that between 38 Ma and 20 Ma there was asymmetric spreading with half-spreading rates of 31 mm yr^{-1} on the north flank and 22 mm yr^{-1} on the south flank. Clearly, sea-floor spreading is not always symmetrical.

Changes of plate motion at ridges

So far, we have only considered situations in which the direction of plate movement at spreading ridges was assumed to be constant over time. But this is rarely the case. As we have seen, a common type of plate boundary consists of straight sections of ocean ridge offset by transform faults. The ridges are commonly orientated perpendicular to the direction of spreading and may have an equal rate of spreading on either side of the ridge.

What happens if the spreading direction changes? Two models have been proposed which could explain what happens. Figure 2.37 shows how a spreading ocean ridge offset by transform faults might adjust by *rotation* to a change in spreading direction. This change in spreading direction will be revealed by the change in orientation of magnetic anomalies. Figure 2.37a shows a ridge spreading symmetrically east–west (indicated by the

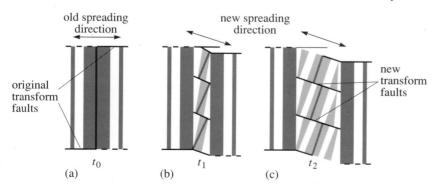

Figure 2.37 A possible model for adjustment of a spreading ridge to a change in spreading direction: (a) at time t_0, symmetrical spreading is occurring; (b) after a change in spreading direction, the ridge breaks into segments, separated by new transform faults, which become realigned at a time t_1; so that (c) symmetrical spreading is re-established by time t_2.

symmetrical magnetic anomalies) at time t_0. Now consider what happens if the spreading direction changes, so that the ridge spreads more towards the northwest and southeast. The ridge then must break into sections, separated by new transform faults. The magnetic stripes formed during the readjustment are not symmetrical, but may approach triangular form (dotted magnetic anomalies in Figure 2.37b, formed at t_1). Finally, as symmetrical spreading continues, the magnetic stripes become parallel again at time t_2 (Figure 2.37c) and a new ridge–transform plate margin is established.

An alternative model proposes the growth of a new spreading centre and its development at the expense of the old ridge. This has been called the **propagating rift** model. Look at Figure 2.38. The old rift, with its dark magnetic stripes, is still active, but is being progressively replaced from the north by a new rift, which produces light-coloured magnetic stripes at right angles to its new spreading direction. Between the new and old rift systems, oceanic crust is being transferred from one plate to the other, which gives rise to a sheared and disrupted zone with a distinctive angular set of magnetic stripes, clearly seen in Figure 2.38. This model therefore predicts abrupt boundaries between areas of otherwise uniform magnetic anomalies, whereas the rotation model predicts a continuous fan-like arrangement of magnetic anomalies. Detailed magnetic surveys should be able to distinguish between the two. If the propagating rift model exists, there should be, somewhere on the global ridge system, a new rift overtaking a pre-existing one and producing a diagnostic set of abrupt changes in magnetic anomaly pattern.

Magnetic surveys were carried out in the late 1970s to test these two models and a propagating ridge segment was discovered in the Cocos Plate. Results from west of the Galápagos Islands show that a new ridge is progressively breaking through the Cocos Plate (Figure 2.39). These magnetic data provide good evidence that the propagating rift mechanism is in operation here.

new spreading direction
old spreading direction

Figure 2.38   Ridge adjustment by rift propagation. New magnetic anomalies are created by the new ridge segment (light stripes) at the same time as new magnetic anomalies are created by the old ridge segment (dark stripes). New and old ridges are both active at the same time, and in between the two ridges oceanic crust is sheared because of the opposite movement direction on the two ridge segments (rather like a transform fault). The sheared oceanic crust recently formed from the old ridge segment carries a distinctive angular pattern of magnetic anomalies.

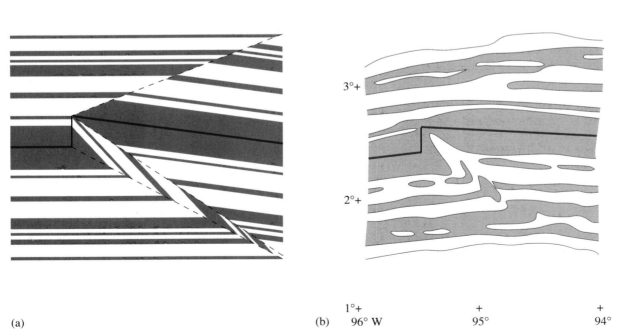

(a)   (b)   96° W   95°   94°

Figure 2.39   Predicted magnetic anomaly pattern (a) from the rift propagation model and (b) observed pattern west of the Galápagos Islands, at 96° W 2° N. The rift propagation model predicts that acute angles should form between magnetic stripes, reflecting changes in spreading direction. Detailed observations from this part of the ocean floor show that such acute angles do exist, and lend weight to the model.

2.4.2 PLATE BOUNDARY CHANGES DUE TO SUBDUCTION

The most profound changes to plate boundaries occur as a result of subduction. We know that all the present oceanic crust is younger than about 170 Ma. Oceanic crust did exist before that time; we have evidence from ophiolites (oceanic crust preserved on the continents) which give radiometric dates older than 170 Ma to confirm this. It follows then that almost all the oceanic crust generated before 170 Ma has now been subducted and destroyed, including any complex triple junctions or ridge segments that may have existed within the plate structure. The nature of any oceanic plate is constantly changing, therefore, as major features formed during its construction are subducted.

Consider the boundaries of the Pacific Plate in Figure 2.40. This is an ingenious map projection of the Pacific, quite different from the Mercator projection used in the Smithsonian Map. The arrows on Figure 2.40 show the direction of motion of the Pacific Plate relative to the North American and Eurasian Plates. This map is drawn so that the direction of motion of the Pacific Plate is parallel to the top and bottom of the map.

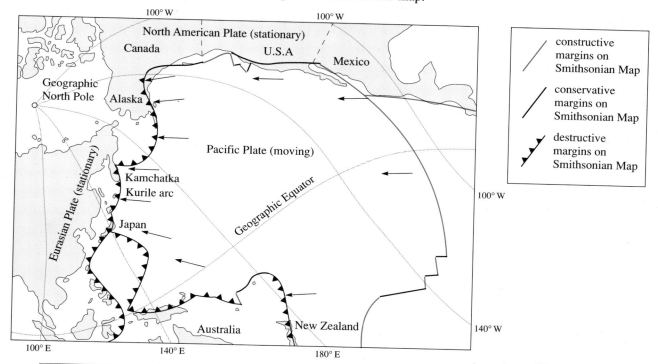

Figure 2.40 A map of the Pacific drawn so that the direction of motion of the Pacific Plate is parallel to the top and bottom of the map. The arrows show the direction of motion of the Pacific Plate relative to the North American and Eurasian Plates. (Note that, although on the scale of this diagram the plate boundary through the Gulf of California appears as a simple spreading ridge, it is, in fact, a spreading ridge highly segmented by transforms.)

ITQ 32

The arrows on Figure 2.40 show the direction of motion of the Pacific Plate to be mainly right to left in relation to other plates (which are assumed to be stationary).

(a) What is the direction in geographic terms?

(b) How does the nature of the American margin of the Pacific Plate change in passing from the right edge of the Figure to the left?

ITQ 33

What will happen to the East Pacific Rise in time?

Perhaps the most significant change that can happen to plate boundaries with time is subduction of an active spreading axis. While we can surmise that this will occur in the Pacific in the next 200 Ma without much fear of contradiction, we cannot say what exactly will happen to the plate boundaries. But by matching existing stripes on the ocean floor and some careful mathematical detective work, we know that subduction has occurred on the northern part of the East Pacific Rise over the past 50 Ma.

Case Study: The Farallon Plate

Look at the northeastern boundary of the Pacific Plate on the Smithsonian Map. We have already studied the East Pacific Rise in some detail around the Clipperton Transform at 105° W 10° N in Section 2.3.2. Following the East Pacific Rise northwards we can trace it across several major transforms into the Gulf of California, where it becomes more transform than ridge, and into the American Plate at the San Andreas Fault zone. The spreading ridge reappears at the Mendocino Transform at 127° W 40° N, and disappears under the North American Plate again at Moresby Island (132° W 53° N). The small fragments of oceanic crust on the American side of this spreading ridge, off Washington State and in the southern part of the Gulf of California, are the remnants of the once-larger Farallon Plate.

About 35 Ma ago, the East Pacific Rise separated the Pacific and Farallon Plates off western USA (Figure 2.41). At this time, the Farallon Plate was being subducted beneath the North American Plate. The rate of subduction exceeded the rate of spreading at the ridge, however, so the ridge system moved towards the trench.

The first part of the ridge to meet the trench was the RF junction at the eastern end of the Mendocino Transform. A quadruple junction (TTRF) existed momentarily about 28 Ma ago, but immediately split into two triple junctions, both stable. The more northerly was of the FFT type while the southerly was of the RTF type. The geometry of the system forced the FFT junction to move northwards along the trench and the RTF junction to move southwards along the trench. A dextral conservative plate boundary fault formed in response to the migration of these triple junctions.

When the Murray Transform reached the trench, the situation changed once again. Another quadruple junction formed and quickly decayed. The southern triple junction changed to the FFT type (still stable) and began to move northwards. The Farallon Plate continued to be subducted to the north and south of the dextral transform fault until the East Pacific Rise once again reached the trench. Yet another quadruple junction formed and decayed, and the geometry changed back to the 28 Ma situation. The southern triple junction reverted to an RTF type and migrated southwards once more along the trench.

At the present time, the southern RTF triple junction has moved southwards to lie in the mouth of the Gulf of California and the dextral conservative plate margin has grown much larger and now forms the well-known San Andreas Fault zone.

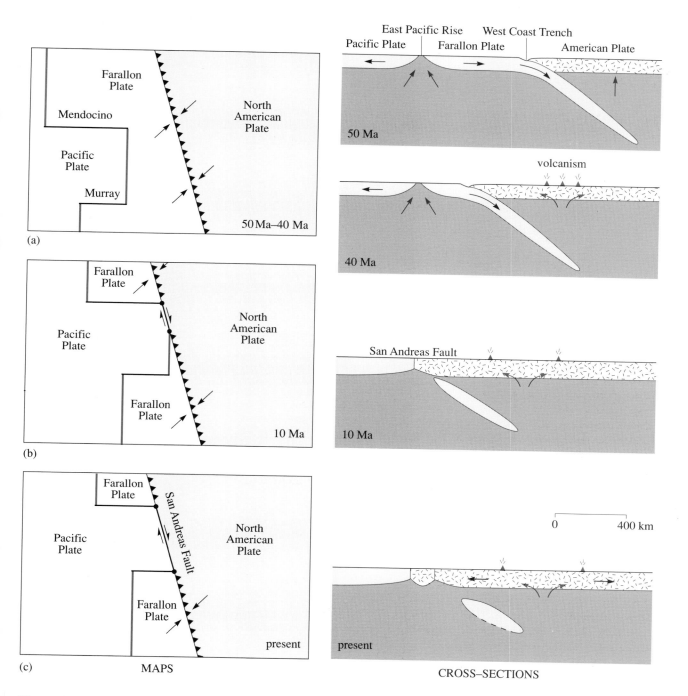

MAPS

CROSS–SECTIONS

Figure 2.41 Subduction of the Farallon Plate and the East Pacific Rise under North America.

SUMMARY OF SECTION 2.4

- Plate construction at spreading axes can be uneven, giving rise to asymmetric magnetic stripes. Often, uneven spreading reflects a change of plate motion at spreading ridges.

- Changes in spreading direction can be achieved either by rotation within an existing rift system or by a new system propagating onto an old one. Examples of both exist in the present-day plate configuration.

- In time, all the features of oceanic plates become subducted. Most significant, as far as changing plate boundaries are concerned, is the subduction of an active spreading axis. This can lead to formation of new boundary types, such as the San Andreas dextral fault zone generated by subduction of the Farallon Plate.

OBJECTIVES FOR SECTION 2.4

When you have completed this Section, you should be able to:

2.1 Recognize and use definitions and applications of each of the terms printed in bold.

2.11 Outline the geological and geophysical evidence for spreading axes evolving with time.

2.12 Describe what happens to the physical features within oceanic plates as they reach a destructive plate boundary.

2.13 Predict, in very general terms, the appearance of a present-day oceanic plate at some stage in the future, based on a knowledge of plate movement directions and plate features.

Apart from Objective 2.1, to which they all relate, the two ITQs in this Section test the Objectives as follows: ITQ 32, Objectives 2.12 and 2.13; ITQ 33, Objectives 2.11 and 2.13.

You should now do the following SAQs, which test other aspects of the Objectives.

SAQS FOR SECTION 2.4

SAQ 7 (*Objectives 2.1, 2.11 and 2.12*)

Based on the information in this Section and in the Smithsonian Map, what is the likely fate of the remnant of the Farallon Plate? (Located around 125° W 45° N.)

SAQ 8 (*Objectives 2.1, 2.12 and 2.13*)

By using the plate movement information on the Smithsonian Map and Figure 2.40, and your answer to SAQ 7, predict the appearance of the Pacific Ocean, its ridges and hot spots and its continental margin 50 Ma into the future.

2.5 PLATE TECTONICS AND CONTINENTAL CRUST

Our discussion concerning plate boundaries and how they change with time has so far concentrated exclusively on oceanic crust, almost as if there were no continental crust at all. From a geoscientific point of view, the continents are arguably more important and very different from the oceans. We live on the continents and we are able to study their geology, geophysics and geochemistry much more easily than we can study oceanic crust. We have been studying the continents for much longer than we have the oceans, too. What then are the differences between continental and oceanic crust and how does this affect the nature of plates and plate boundaries?

2.5.1 HOW CONTINENTAL CRUST DIFFERS FROM OCEANIC CRUST

You may recall from Section 2.3 and from Block 1 that continental and oceanic crust differ considerably. These differences are responsible for the profound differences in behaviour between oceanic and continental plates. For our purposes, we shall loosely define **oceanic plates** as plates which contain primarily oceanic crust and mantle in their lithosphere (e.g. Pacific Plate) and **continental plates** as plates that have significant components of continental crust (e.g. North American Plate).

❑   Referring back to Section 2.2 and Block 1 as necessary, name three differences between the physical and chemical characteristics of continental and oceanic crust.

■   There are many differences in physical and chemical features between the two and you might have suggested any from the following list (or even some others). The main differences are:

- P-wave velocities in oceanic crust are typically $4.5–5.5\,\mathrm{km\,s^{-1}}$ for layer 2 and $6.5–7.0\,\mathrm{km\,s^{-1}}$ for layer 3. Contrastingly, P-wave velocities in the lower continental crust range from $6.5–7.6\,\mathrm{km\,s^{-1}}$. There is no simple relationship between P-wave velocity and depth.

- The seismic Moho lies at around 7 km beneath the ocean floor but underneath continents it is typically 40–50 km deep.

- The average density of oceanic crust is between $2\,600\,\mathrm{kg\,m^{-3}}$ (layer 2) and $2\,900\,\mathrm{kg\,m^{-3}}$ (layer 3). The average density of the upper continental crust is about $2\,650–2\,850\,\mathrm{kg\,m^{-3}}$ while the lower continental crust is about $2\,900–3\,000\,\mathrm{kg\,m^{-3}}$. The bulk of oceanic crust is therefore denser than the upper parts of the continental crust. On average, continental crust is less dense than oceanic crust.

- Oceanic crust is almost wholly of silica-poor basaltic composition. By contrast, continental crust is mainly composed of silica-rich minerals. The mean composition of the upper crust corresponds to the rock type granodiorite in composition. Lower continental crust is typically rather less silica-rich.

- Radiogenic heat production is greater within continental crust than within oceanic crust.

- Almost all oceanic crust is less than 170 Ma old, whereas continental crust is typically much older than this. The oldest continental crust is over 3 600 Ma old.

- Oceanic crust is constantly being recycled whereas the continents have been joined and rejoined several times during Earth's history.

These properties allow us to draw some general conclusions about the different behaviour between continental and oceanic plates.

First, both continental and oceanic crust are less dense than mantle material. Both will therefore 'float' on rocks of normal mantle composition. However, the upper continental crust in particular is considerably less dense than either mantle material or oceanic crust. This means that normally the continents do not get subducted. Whilst subduction of oceanic crust may not be seriously hampered by any tendency of the downgoing material to be buoyed up by isostatic forces, the same is not true of upper continental crust.

Secondly, this conclusion is supported by the age of continental crust. If continental crust were to be constantly recycled by subduction, we would not expect to find any continental crust substantially older than the oldest oceanic crust. Yet the bulk of the continental land area is composed of rocks older than 700 Ma, as you can see in Plate 2.5. Rocks of this age are over three times as old as the oldest known oceanic crust. Only a very small percentage of continental rocks (those coloured mauve in Plate 2.5) are as young as the present oceanic crust.

Thirdly, the fact that the continents have split apart and 'wandered' over the globe throughout the Earth's past suggests that they have experienced a form of plate tectonics which is very different from the oceanic plate cycle. The continental cycle appears to involve splitting–drifting apart–drifting together–rejoining, rather than the oceanic crust cycle of formation–transport–destruction.

We don't have the opportunity in this Course to study continental tectonics in much detail, so we shall concentrate on the continental parallels of the major oceanic plate boundaries. Each will be illustrated with a case study and the most important aspects of continental plate tectonics will be summarized at the end of this Section.

2.5.2 CONSTRUCTIVE MARGINS WITHIN CONTINENTAL CRUST

At oceanic constructive margins, new oceanic crust is forming and being added to the edges of oceanic plates. Does something similar happen at constructive margins within continents? To investigate this point, let us first remind ourselves of the processes at constructive oceanic plate margins and see which of these processes have parallels in the continents.

At constructive oceanic boundaries, a cyclical process takes place that is governed by **extension**. We know that extension is taking place here by examining the movement directions of the two rock bodies that are displaced during earthquakes. At spreading ridges, oceanic crust and its underlying mantle is being stretched by the forces that are pulling the two plates apart. The main extensional forces are roughly horizontally oriented.

Oceanic lithosphere, when stretched near spreading ridges, deforms in a dual way. The uppermost parts of the newly built plates fracture; we see the evidence for this in the sheeted dykes which form part of layer 2 of oceanic crust. This part of oceanic crust is cool because it is subjected to the chilling effect of circulating seawater. Since it is colder it behaves in a brittle manner, like glass. The lower part is ductile, so the mantle stretches and thins, allowing the asthenosphere to rise and start to melt.

Continental crust is much thicker than oceanic crust. The Moho, which marks the base of the crust, lies at about 7 km under oceanic crust but at an average 40–50 km under the continents. Plate boundaries within continental crust become broad and diffuse compared with the narrow and well-defined oceanic boundaries. Also, the continental crust is composed of irregular areas of rocks of different compositions and different ages, so it is much less uniform in its physical properties than is oceanic crust. In particular, old tectonic lines, major fault zones or possible former plate margins will have a great influence on the exact way continental crust responds to stretching forces. Together, these two points mean that a broad zone will possibly act as a plate boundary, rather than the fairly narrow, discrete zones found in the oceans.

If a constructive plate margin (which we might more usefully call a divergent plate margin in this context) passes into or forms within continental crust, the crust and its underlying mantle will stretch. The upper parts of the crust stretch and fail in a brittle manner, developing faults that have an extensional character. These faults develop in long, narrow zones and within these zones the crustal rocks subside and tilt differentially, forming terraces of tilted and rotated fault blocks (Figures 2.42 and 2.43). Such features are called **rift valleys** or **grabens** (German: *Graben*, ditch or trench). Each of the faults in a rift valley allows stretching to take place by extending the length of pre-rift sedimentary layers.

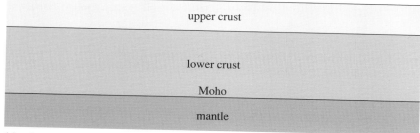

(a) prior to extension

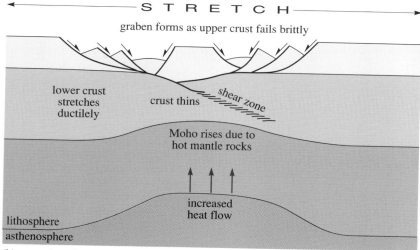

(b) extension phase

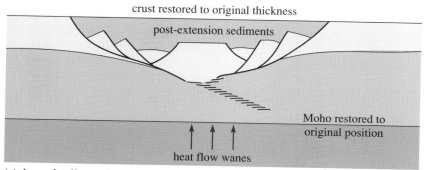

(c) thermal collapse phase

Figure 2.42 Divergent (constructive) margins within the continents.

If stretching amounts are modest, extension is achieved both by block rotation and normal faulting and surface subsidence is relatively slight. The process can't halt here, because the whole area is still in thermal disequilibrium and the lower surface of the lithosphere is still elevated (Figure 2.42b). As the thermal anomaly decays and fades away, the elevated lithosphere will subside, allowing unfaulted sediments to accumulate in the basin so created (Figure 2.42c). There are many examples of this type of 'failed' divergent boundary, characterized by normally faulted pre-rift sediments (with or without some volcanic rocks) buried by a wedge of unfaulted sediment. The best example close to home is the sedimentary basin of the North Sea (Figure 2.43).

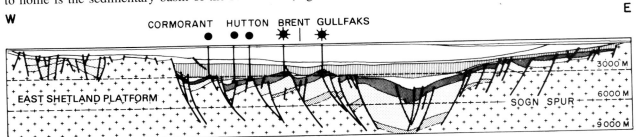

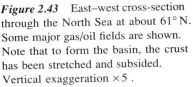

Figure 2.43   East–west cross-section through the North Sea at about 61° N. Some major gas/oil fields are shown. Note that to form the basin, the crust has been stretched and subsided. Vertical exaggeration ×5 .

In other cases, divergence continues and the continental lithosphere becomes more and more thinned. The rising asthenosphere below occupies an area of reduced pressure as its cover is removed by thinning, and starts to melt. Volcanic rocks become increasingly important within the basin. The whole process ends in the separation of two bodies of continental crust and the formation of a new zone of oceanic crust in between (Figure 2.44). Each continental margin moves away from the new spreading axis, cooling and subsiding as it does so. The normal faults that helped the crust to stretch initially become inactive, and the continental margin becomes blanketed by coastal sedimentary rocks. The edge of the continental crust now has all the characteristics of a passive continental margin.

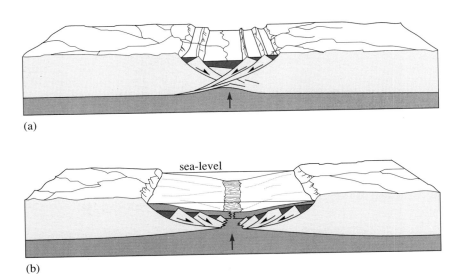

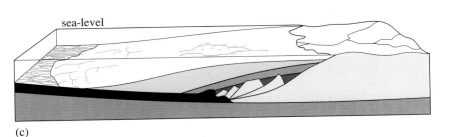

Figure 2.44   Cross-section of the sequence of events at a divergent continental plate boundary. Two plates move apart, causing extension and thinning of the continental crust. At first (a), the upper parts of the crust extend by developing a series of brittle normal faults. Often, the normal faults rotate as the crust subsides. The continental surface also sinks, creating a basin which provides a site for accumulation of sedimentary rocks or volcanics (darker tone). Extension at deeper levels is accommodated by ductile processes; either stretching or the ductile equivalent of faults called shear zones. As the plates continue to diverge (b), the lower lithosphere also stretches, allowing the asthenosphere to rise, decompress and melt. Eventually (c), oceanic crust forms in the centre of the basin and the continent has separated into two. The thinned margin of each subsides as it cools and moves away from the spreading site. This subsidence allows a further cover of unfaulted sediments to be deposited. A passive margin has been created.

Occasionally, a continental divergent boundary is located on the continental side of an older continental margin, which may itself be a divergent boundary. Two continental blocks would then exist which might be quite unequal in size, and fragments of a continent (often called **microcontinents**) may be rafted away to form small plateaus of continental material surrounded completely by oceanic crust. A good example of a microcontinent, which is actually forming at the present day, is the Baja California peninsula on the Pacific coast of Mexico and the USA. This thin sliver of continental crust is being separated from mainland North America by the San Andreas Fault system and the spreading ridge which goes through the Gulf of California. Over time, in the next 20–50 Ma, Baja California will become a continental crustal island surrounded by oceanic crust.

Case Study: The Gulf of Aden

The position of the Gulf of Aden in relation to the plate boundaries that surround the African continental plate is clear from Figure 2.45. Together with the Red Sea, it forms a V-shaped rift that separates the African from the Arabian continent.

Figure 2.45   Major plate boundaries and tectonic features of the African Plate.

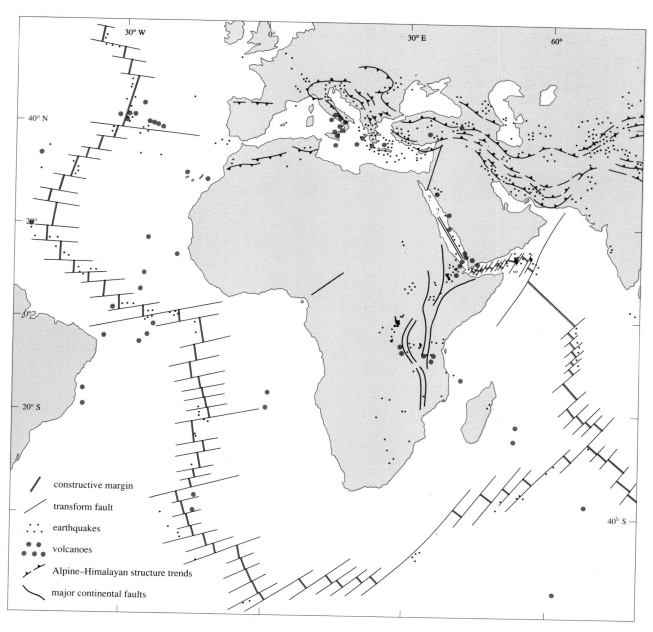

/	constructive margin
/	transform fault
∴	earthquakes
●	volcanoes
⌁	Alpine–Himalayan structure trends
⌇	major continental faults

The Gulf of Aden is about six times as long as it is wide, and the Red Sea is ten times longer than it is wide. The shapes alone suggest that they are young plate tectonic features. A mature ocean would be broader: the Atlantic is twice as long as it is wide. Both the Red Sea and the Gulf of Aden have spreading axes centrally placed along their length, but as Figure 2.46 shows, the two seaways have contrasting spreading patterns. The Red Sea has one long spreading axis positioned centrally and parallel to its length with no obvious transforms. The Gulf of Aden, however, has a segmented spreading axis offset by transforms. This difference occurs simply because the Red Sea is spreading almost at right angles to its length, whereas the Gulf of Aden is spreading oblique to its length.

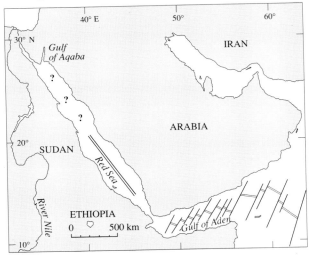

Figure 2.46   Spreading axes and transform faults within the Red Sea and Gulf of Aden.

Figure 2.47 records the development of the Gulf of Aden over the past 10 Ma. Figure 2.47a shows the Gulf as it is today. The central red zones represent ocean-floor ages, with the youngest (darkest red) immediately along the present spreading axis. This diagram also shows that the southern end of the Red Sea contains a thin strip of young oceanic crust. By removing these stripes sequentially from such a diagram, geoscientists can effectively restore the Gulf of Aden to its pre-rifting positions and explore how continental separation and spreading have taken place.

Figure 2.47b shows how the Gulf of Aden looked some 7 Ma ago. Comparing this with Figure 2.47a shows two interesting features. First, the Gulf is clearly spreading obliquely. Note the relative movement of the Ras Asir peninsula on the northeast tip of Africa away from the Ras Fartak peninsula in Arabia. Spreading is in fact parallel to the transform faults. Secondly, the rift is clearly propagating westwards. The tip of the rift permitting new ocean floor to be created has moved from off the African continent (b) to its present location in the Afar depression (a), some 200 km further west.

ITQ 34

How fast is the spreading tip propagating westwards, in mm yr^{-1}?

Figure 2.47c brings out a third important point about continental separation. In this diagram, showing plate positions 10 Ma ago, there is a clear overlap between the reconstructed positions of the African and Arabian continental crust.

How can these overlaps exist? They could be due to formation of new features after rifting, like we saw in the case of the Niger delta in ITQ 1. This is not appropriate here, because both areas of overlapping continental crust are largely composed of granitic and metamorphic rocks some 500–1 200 Ma old. Old continental crust can't have been generated since rifting

74

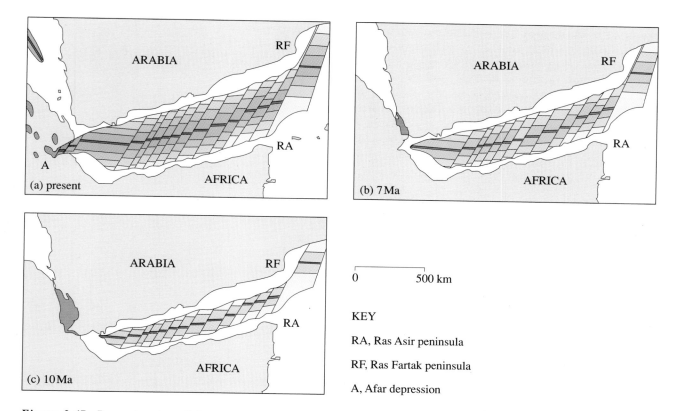

Figure 2.47 Reconstruction of the separation of the Gulf of Aden over the past 10 Ma. (a) Present-day plate position: Magnetic stripes formed by recently erupted oceanic crust are shown in red (darker red for younger strips). (b) Plate position 7 Ma ago; and (c) 10 Ma ago. Over the past 10 Ma, Arabia has moved NNE away from Africa, relatively, and the tip of ocean-floor generation has moved west into Africa. Note the areas of apparent overlap of continental crust in (b) and (c). These show that the plates have stretched as they rifted.

began. In fact, both areas also contain basalts less than 30 Ma old, and the overlap is caused through stretching of the pre-existing continental crust. The amount of overlap is a measure of the amount of stretching of the continental material prior to continental separation and sea-floor spreading. In settings like this, continental crust has significantly increased in area in the recent geological past.

In summary, the Gulf of Aden and the Red Sea form a linked stretching system by which Arabia is rifting away from Africa. A once-continuous continent is breaking into two and a new ocean basin is starting to form. New oceanic crust is presently being created in the centre of the Red Sea and has been present for over 10 Ma along the axis of the Gulf of Aden. Crustal stretching started at least 30 Ma ago, since the ancient continental crust of the Yemen peninsula of Arabia and the Afar area of Africa has stretched by faulting and been intruded by basaltic magmas which are up to 30 Ma old.

2.5.3 DESTRUCTIVE MARGINS AT THE OCEAN–CONTINENT BOUNDARY

At oceanic destructive margins, old oceanic crust is being destroyed beneath an overriding oceanic plate. Is the process different if continental crust is involved? To investigate this point, we will follow the pattern established earlier in this Section — we'll first outline the principles of destructive oceanic plate margins and then see which of these principles have parallels in the continental setting.

We saw in the Lesser Antilles subduction zone that the volume of continental crust was being increased at an oceanic subduction zone in two ways. First, plutons derived from the mantle were being emplaced into the crust in the root of the active island arc, and lavas were also being erupted at the surface. Both lavas and plutons are more silica-rich than oceanic crust and closer in composition to continental crust, as we will discuss further in Block 4. This process is adding to the crust by emplacing material derived from the mantle. Secondly, sediments deposited onto the subducting plate in the deep ocean are accreted onto the overriding plate. This second process is not generating new continental crust exactly, rather it is recycling pre-existing material. It might be better to call this process relocation of continental crust.

The tectonic processes at a convergent margin, even when the two plates involved are composed essentially of oceanic crust, can be considered to be the opposite of those going on at a divergent margin. Whilst the stresses are extensional at divergent margins, at a convergent margin **compression** is the order of the day. We know that compression is taking place here by examining the movement directions of rock bodies displaced at subduction zones during earthquakes. The convergent movements of the two plates compresses the accretionary prism, forming both faults and folds in the accretionary wedge (Figure 2.48).

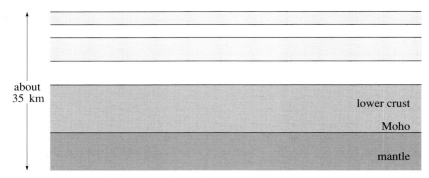

(a) Original layered continental crust

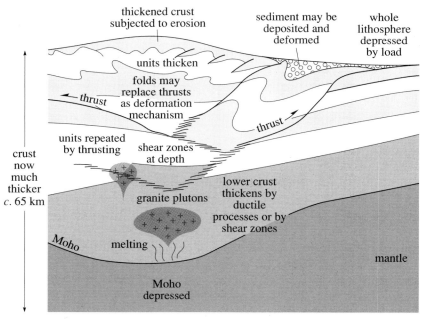

(b) Crust that is shortened thickens by folding and thrusting

Figure 2.48   How thrusts and folds compress and shorten rocks.

In compressional zones, the main compressional forces arise from the horizontal movements of plates, so they too are roughly horizontally oriented. The lithosphere behaves in a twofold brittle–ductile manner, as it does under extension. Under horizontal compression, ductile materials thicken vertically and shorten horizontally, thickening the plate. The brittle upper parts of the plate, and the cold sediments in the accretionary wedge, commonly fracture along failure planes or faults. Faults that shorten rocks are called **thrusts**. Under certain conditions, folds accommodate some of the crustal shortening. Both folds and thrusts achieve the same end — shortening and thickening crust (Figure 2.48).

How does the presence of continental crust modify this type of plate boundary? There are several new situations that might occur at destructive margins if continental crust is involved in the plate configuration. There may be continental crust in the overriding plate, but not the subducting plate; or vice versa with continental crust in the subducting plate but not in the overriding plate; or there may be continental crust in both plates. Continental crust may lie at the margins of either plate, or there may be a large area of oceanic crust between the subducting margin and the nearest continent. There may be equal amounts of continental crust on both plates or, more likely, very much more continental material on one than on the other. The situation is potentially very complex.

To simplify our study, we will consider two important geometries that are part of the present cycle of plate tectonics. The first may perhaps be considered the 'natural' continental margin, where an oceanic plate is subducted beneath continental crust, frequently called an **Andean margin**.

Case Study: The Andes

On the west coast of South America, the oceanic Nazca Plate is being subducted beneath the South American continent. The two plates are moving almost exactly towards each other, as can be seen from the movement vectors on the Smithsonian Map.

As the Nazca Plate subducts beneath the Andean margin, magma is produced, either from the oceanic plate itself or from the South American mantle wedge immediately above it. The magma may erupt to form volcanoes, or it may solidify within the continental crust to form plutons. This part of the process is almost identical to that happening at oceanic plate boundaries, so we can tell that this part of the process must be independent of the continental crust. However, the major part of the upper continental crust in the Andes is not lavas or plutons, but sedimentary rocks formed in a coastal or marine environment. These sediments were deposited on the flanks of South America, possibly during rifting or thermal subsidence phases of the spreading part of an earlier cycle.

How do these submarine sediments now come to form the highest mountains in the Western Hemisphere? This is because they are now buoyed up by a **crustal root** that seismic studies tell us is in places more than 70 km below the current land surface (Figure 2.49). Why is the continental crust so thick here? We have already mentioned one reason; magmas derived originally from the mantle have been emplaced into pre-existing continental crust. Estimates made by geoscientists working in Peru suggest that the emplacement of granitic plutons in this part of the continental crust alone has increased its width by over 200 km in the past 200 Ma. This must have been accompanied by significant upper crustal thickening, the exact amount depending on the exact shape of the plutons. However, all of this plutonic material does not come from the mantle; some is derived from melting of lower crustal rocks, as we shall see in Block 4.

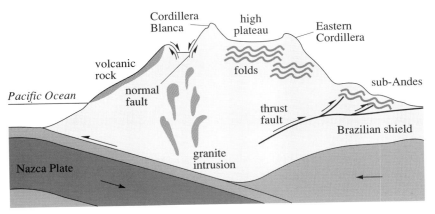

Figure 2.49 A diagrammatic view of how the Peruvian Andes are buoyed up by a thick continental root and by flexural support from the South American Plate. Under the Western Cordilleras, the crust has been thickened by intrusions of magma derived from the subduction of the Nazca Plate, though mostly not from the plate itself but from the overlying mantle wedge. The convergence of the two plates also compresses the pre-existing continental crust and thickens it. Although the sides of the Andes are still being pushed together, parts of the range are extending as the mountain belt collapses.

The continental crust is also tectonically thickened by the development of thrusts and folds. Both of these structures allow the crust to shorten horizontally and thicken vertically. The orientation of both types of structure tells us that shortening took place approximately perpendicular to the width of the mountain belt (Figure 2.50). As folds and thrusts essentially conserve the volume of rock involved, they must also either thicken the crust as a whole or lengthen the zone. Ultimately, it proves more difficult to lengthen the mountain belt against the resistance of the surrounding continental plates, so thickening normally takes place.

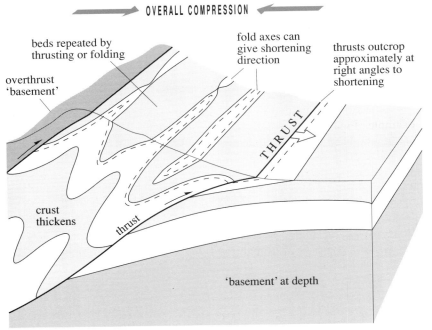

Figure 2.50 The orientation of thrusts and folds gives generalized information about the direction of shortening of a mountain belt.

Evidence for brittle displacement of the Andean zone can be found in the thrust structures of the sub-Andes (Figure 2.49). Seismic studies in this area show that earthquakes are focused at depth on the eastern flanks of the Andes as the continent of South America is being thrust westwards under the mountains. The rate of underthrusting, at only several mm yr^{-1}, is much slower than oceanic plate movement.

2.5.4 DESTRUCTIVE MARGINS AND CONTINENTAL COLLISION

The Andean example described above shows that continental crust thickens and deforms above subduction zones by both igneous and tectonic processes. The driving force for both processes is the convergence of the plates. As subduction is a more or less continuous process, we might expect that growth of the continental crust will be more or less continuous, too. In the Andes, a mountain belt has formed by this process over the past 200 Ma.

However, a quick glance at Plate 2.4 shows that mountain belts don't always lie near to active continental margins. Indeed, the highest mountain belt in the world at present, the Himalayas, is completely surrounded by continental crust, separated from the nearest ocean by the continent of India. Many of the other continents contain internal mountain belts which are similarly distant from present-day oceans; the Urals of central Russia is a good example.

What plate tectonic process might generate mountain ranges in the middle of continents? The answer is not obvious, particularly as it involves a plate tectonic process that does not at first appear to be happening at the present day. Three possibilities exist. Mountain belts might have formed where they are now found, within continental crust and surrounded by continents. Alternatively, mountain belts might be generated at a continental margin, and if so, some process must exist that allows continental crust to accrete onto the outer margins of an existing continent. Further, as we know continents drift with time, mountain belts may possibly form when continents drift towards each other and ultimately collide into each other.

ITQ 35

How could geoscientists investigate which one or more of the three possibilities is correct?

Case Study: The Himalayas

Let us examine which of the three proposals — *in situ* growth of mountains within an existing continent, *in situ* growth at the edge of a continent, or growth of mountain belts due to continental collision — applies to a modern mountain belt using the approach highlighted in ITQ 35. We shall use the Himalayas as an example. Figure 2.51 shows the positions of the southern continental land-masses 115 Ma ago, based on palaeomagnetic evidence. You can see that not only was South America joined to Africa at that time, but all the land-masses of the Southern Hemisphere were once joined to form one huge continent called **Gondwanaland** by geoscientists. We know from both polar wander and oceanic magnetic stripes that India was part of the continent of Gondwanaland, too.

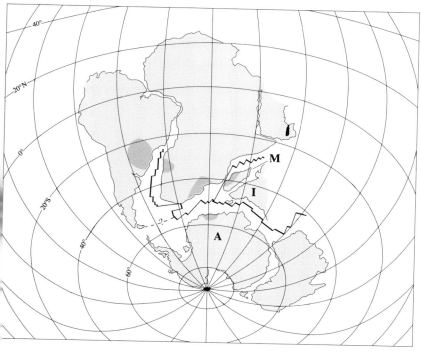

Figure 2.51 Reconstruction of the southern continents at about 115 Ma ago.
I is India, M is Madagascar and A is Antarctica. The red line shows the spreading axes responsible for breaking up the existing continent. Note India's position, determined from palaeomagnetic evidence, at about 40° S, close to Antarctica.

In the period between 115 Ma and the present, India broke free from Gondwanaland and drifted northward. We know this both from palaeomagnetic studies of continental material and by plotting former positions of the Indian subcontinent using ocean floor magnetic stripes (like we did for the Gulf of Aden). These studies prove that India impacted into Asia between 55 Ma and 40 Ma ago. Figure 2.52 charts part of this northward drift.

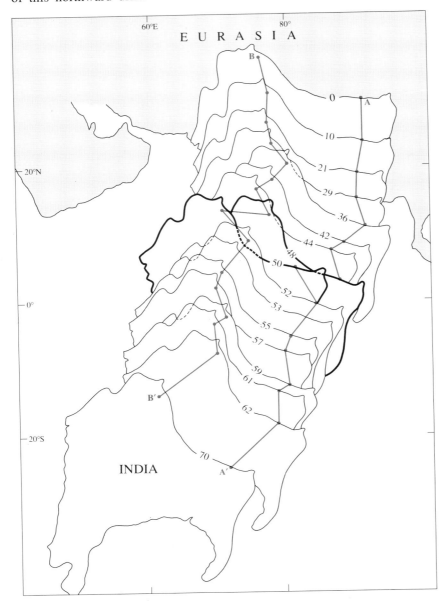

Figure 2.52 The northward drift of India relative to Asia over the past 70 Ma. The numbers on the northern margin of the Indian continent indicate the age (in Ma) when it was in that relative position. In this diagram, Asia is kept fixed arbitrarily.

ITQ 36

Using the latitude data on Figures 2.51 and 2.52, what is the minimum rate of relative convergence between India and Asia over the past 115 Ma? (*Hint*: repeat the technique you used in ITQ 2, using the same Earth radius of 6 370 km.)

Dating of the igneous rocks that intrude the Himalayan mountain chain shows that the plutons within the mountain belt were formed from subduction-related magmatism 110–40 Ma ago. This corresponds almost exactly to the period when India was tracking northward before eventually colliding with Asia. The almost inescapable conclusion is that the Himalayas were formed from a combination of Andean-type processes between 110 and about 50 Ma ago, and by continental collision between about 50 Ma and the present.

Oceanic crust which originally lay seawards of a passive margin on the Indian Plate started to be subducted beneath Tibet over 100 Ma ago. An Andean-type margin was established on the southern margins of Tibet which persisted for over 60 Ma (Figure 2.53a).

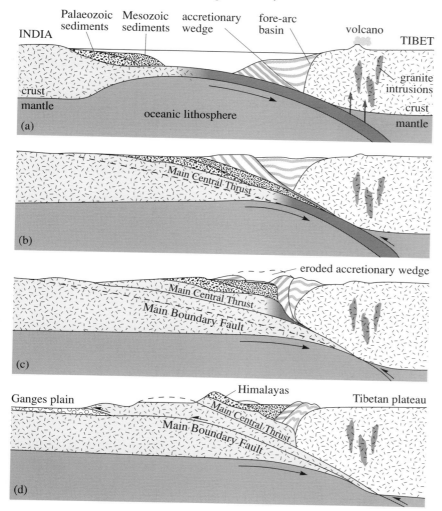

Figure 2.53 The formation of the Himalayas depicted schematically. Andean-type processes between 110 and about 50 Ma ago, and continental collision between about 50 Ma and the present, together have created this mountain chain.

Between 55 Ma and 40 Ma ago, the two continents began to collide (Figure 2.53b). The Indian continent was less dense than the mantle beneath Tibet and therefore too buoyant to be subducted. The Main Central Thrust was generated, as a brittle fault plane in the upper crust and as a ductile thrust zone in the lower parts of the crust, which effectively allowed the Indian continent to shorten and thicken. The Main Central Thrust became inactive as it was lifted to upper crustal levels about 20 Ma ago (Figure 2.53c) and its role in allowing the whole Indian continental plate to shorten was taken over by the lower Main Boundary Fault. Thrusting continues today at high crustal levels as far south as the Ganges plain.

The Himalayas are the highest mountains in the world because they are buoyed up by anomalously thick continental crust — the Moho under the High Himalayas lies at more than 70 km. This crustal thickening at all levels has been achieved by folding and thrusting (Figure 2.54).

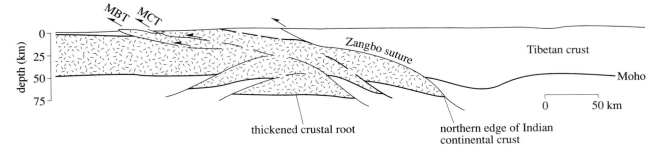

Figure 2.54 A true-scale cross-section through the whole continental crust under the Himalayas, based on an interpretation of seismic refraction data. Notice how thrusts can shorten the crust at all levels, from the surface right through to include the Moho at over 70 km.

Burial of continental crust leads to rises in temperature and pressure, which themselves produce changes in the character of the original rocks. This change may include an alteration in the minerals that make up the original rock. Usually, minerals with a new chemical composition grow, which are physically and chemically stable under the new temperature and pressure conditions. These growing minerals will also have an alignment which reflects the deformation processes influenced primarily by pressure. Rock types such as slates, schists and gneisses are characteristic of these changes. The process as a whole is known as **metamorphism** (from the Greek for 'changed shape'), and the new rock types with their distinctive texture and mineralogy are known as **metamorphic rocks**. Metamorphic rocks form a major component of areas of convergent tectonism.

So, to sum up, mountain belts form at active continental margins, either as a result of Andean-type processes, where oceanic plates are subducted, or as a result of continental collision, or by both, as is the case in the Himalayas. Whatever the detailed plate setting, continental crust thickens by folding, thrusting and intrusion of magma.

2.5.5 CONSERVATIVE MARGINS WITHIN CONTINENTAL CRUST

So far in this Section, we have considered processes taking place in continental crust that are parallels of oceanic spreading axes and oceanic destructive margins. We know that a third type of plate boundary exists in oceans.

ITQ 37

What is that third type of plate boundary and what type of plate motion characterizes it?

Is there a counterpart of the transform fault within continental crust? Deductive reasoning suggests there must be, for as transforms link displaced spreading centres in an oceanic context, so they might also do in a continental setting. There is also the possibility that faults with lateral (rather than vertical) displacements link areas of continental splitting with areas of continental collision, or at least with destructive margins.

Look at the Smithsonian Map to answer these questions:

Where within the continents might we find examples of the two situations described above? Locate one example of a fault zone within continental rocks which links spreading axes together, and one example of a fault zone within continental rocks that links a constructive margin to a destructive one. What lines of evidence that you now know about might be used to show whether plates move or not?

To find examples of continental transforms, we need to locate all the Earth's spreading axes and trace them onshore. There are two good, clear examples. In the East Pacific, the East Pacific Rise can be traced northwards from the Equator almost to the coast of Latin America. From here it is stepped leftwards by many ocean transforms towards and up the Gulf of California. The spreading centre clearly intersects continental crust in southern California at the south end of the San Andreas Fault zone. At the northern end of the San Andreas Fault zone, the spreading axis passes from continental crust into oceanic crust again at about 40° N 125° W. The San Andreas Fault zone appears to link spreading axes within continental crust.

The other place where a spreading axis (as opposed to an oceanic transform) meets continental crust is at the northern end of the Red Sea. Here, in the Gulf of Aqaba, the spreading axis intersects the Dead Sea Fault zone. This fault zone can be traced north into Turkey where it meets the Taurus Mountains, which are the western continuation of the Zagros fold belt in Iran. The continental block of Arabia is moving northeastwards at the present day, enabled by the spreading axis of the Gulf of Aden and the Red Sea to its south, the sinistral Dead Sea Fault zone to its west and the convergent margins of the Zagros and Anatolia to its north and northeast.

To investigate the nature of these continental transforms in detail, we will use the San Andreas Fault zone as an example.

Video Case Study: The San Andreas Fault zone

You should now view VB 03 'Plates in Motion: The San Andreas Fault'. Read the accompanying video notes carefully before viewing. After you have viewed the whole video, tackle the questions at the end of the notes to check your understanding of the concepts that were developed.

2.5.6 EXTENSION WORK

Section 2.5 has touched all too briefly on the broad subject of continental plate tectonics, and much has necessarily been left out. However the Course Team feels you might like to know a little about some of the other plate tectonic situations that exist in the present-day plate tectonic cycle, even though there is no opportunity here to go into any detail. The list that follows is simply to complete the current plate tectonic picture and perhaps to whet your appetite for some independent research in the scientific literature, if you ever get any spare time!

1   There appears to be an immense variety of types of continental collision. The Himalayas represented a head-on continental impact, but in places continents collide obliquely, e.g. the margins of the Indian continent in the Sulaiman and Kirthar Ranges of western Pakistan.

2   Oblique displacement of continental segments, like Baja California on the western side of the San Andreas, must inevitably be followed by **accretion** of that continental material onto the margins of continents. The whole of the western seaboard of North America, north of the San Andreas Fault, is composed of continental slivers that have crossed part of the Pacific and accreted onto the American continent. Because they don't properly belong where they are now found, these are known as **exotic terranes**.

3   **Arc–continent collisions** will often precede full continent–continent collisions, as continents will often have an island arc on their oceanward side. A good example of this process took place during the building of the Western Himalayas, where the Kohistan island arc (now geographically in northern India) accreted onto the Asian Plate about 70 Ma ago.

4   Frequently, extension within continental crust is associated with convergent margins. Good present-day examples include the Basin and

Range Province of western USA and the Aegean region of the eastern Mediterranean.

5   Extension in this setting can rift continental fragments away from the main mass. The Lord Howe Ridge between Australia and New Zealand, and the Japanese islands are good modern examples. Under appropriate circumstances, these fragments may cross oceans and accrete onto new continents as exotic terranes!

6   Continental transforms can link subduction zones too, like the New Zealand Alpine Fault.

7   Since we know that some continental crust is over 20 times as old as the oldest oceanic crust, a fascinating possibility is raised that plate tectonic cycles from the distant past might be recorded within continental mountain belts. We shall return to this speculation in Block 5.

AV 06 'The Smithsonian Map, Part II', serves the dual purpose of reviewing the geophysical evidence for plate tectonics and revising the key points of the whole of Block 2 so far. You will find it useful to listen to it now (running time 23 minutes), and to play it again at the end of the Course to help your revision. (*Note*: There are no AV notes with AV 06.) You will also need to refer to the Smithsonian Map while listening to it.

SUMMARY OF SECTION 2.5

- Continental crust is significantly physically different from oceanic crust. Continental crust shows no simple relationship of P-wave velocity with depth and generally the continental Moho is at least twice as deep as the oceanic Moho. On average, continental crust is less dense than oceanic crust.

- Continental crust is too light to be subducted.

- The continents are much older than the ocean floor. Whilst the oldest crust in the present-day oceans is not older than 170 Ma, almost all continental material is older than that.

- Upper parts of the crust deform in a brittle way by forming faults. At upper to intermediate levels, folds are more characteristic. The lower parts of the crust deform by ductile processes, such as shear zones.

- At continental constructive margins, continental crust deforms by extension. The whole lithosphere is thinned by stretching, so heat flow increases and doming occurs. As the stretch continues, brittle, upper crustal levels develop sets of extension or normal faults, often generating fault-controlled valleys or graben. These valleys are good sites for sediment to accumulate, so constructive margins typically have thick sedimentary sequences.

- As the continental crust stops stretching, either by stress relaxation or by ocean formation, heat flow wanes and thermal relaxation takes place, which in turn produces subsidence. Thick unfaulted layers of sediment characterize this phase.

- Although continental separation could and has occurred in the centres of continents, spreading is often located near the margins of continental blocks. Small volumes of continental crust, or microcontinents, are detached from the main continent. Modern examples include the Arabian microplate and Baja California.

- At continental destructive margins, the crust deforms by compression. Thrust faults and folds thicken the crust of the overriding plate, and magma derived from the subduction process is emplaced into the overriding plate, adding to the thickening.

- When the subducting plate includes continental crust, continental collision inevitably occurs. This may be head-on, like the Himalayas, or oblique, like the Western Cordilleras of North America.

- When collision occurs, the subducting plate is also thickened and shortened by folding and thrusting. The crust is heated because work is done to deform rocks. New minerals grow and the pre-existing igneous or sedimentary rocks are changed into metamorphic rocks.

- Conservative continental margins are characterized by extensive strike–slip fault systems. These displace continental blocks sideways and produce complicated patterns of folding and faulting.

- Conservative continental margins can be produced by the impact of an oceanic spreading system on a continent; the San Andreas Fault zone is a good example. They can also be produced along the margins of spreading plates, like the Dead Sea Rift zone.

- Present-day continental plate tectonics is extremely complicated!

OBJECTIVES FOR SECTION 2.5

When you have completed this Section, you should be able to:

2.1 Recognize and use definitions and applications of each of the terms printed in bold.

2.14 Describe how continental crust differs from oceanic crust.

2.15 Compare the processes at oceanic margins with those at continental margins.

2.16 Account for the different deformation patterns seen at constructive, destructive and conservative continental margins.

Apart from Objective 2.1, to which they all relate, the four ITQs in this Section test the Objectives as follows: ITQ 34, Objectives 2.15 and 2.16; ITQ 36, Objectives 2.2 and 2.15; ITQ 37, Objective 2.16.

You should now do the following SAQs, which test other aspects of the Objectives.

SAQS FOR SECTION 2.5

SAQ 9 (*Objectives 2.1, 2.14 and 2.15*)

Contrast the plate tectonic cycle for oceanic crust with that for continental crust. Why are these two cycles so different? (*About 5 or 6 sentences*)

SAQ 10 (*Objectives 2.1 and 2.16*)

The East African Rift (around 35° E 0° N/S on the Smithsonian Map) is an area where continental crust is currently stretching. Assuming that this process continues to continental separation, describe the likely sequence of tectonic and volcanic events that will take place in this part of Africa.

SAQ 11 (*Objectives 2.1 and 2.16*)

The Scottish Highlands are believed to form the root of a 'fossil' continental collision zone. What features would you advise a visiting team of geoscientists to look for in order to confirm this theory? (*About 150 words*)

2.6 WHAT DRIVES THE PLATES?

One topic remains to conclude our discussion of how plate tectonics works. Why do the plates move and what drives them?

We know already that the forces that drive plate tectonics must be generated in the upper parts of the Earth, because we have seen in Section 2.2 that the plates pass over static hot spots that originate from deep in the mantle. What exactly is the nature of these driving forces? Are the plates pushed at spreading ridges or are they pulled at subduction zones? Alternatively, are they propelled along by some type of convection current in the asthenosphere itself? Possibly more than one of these forces exists. If so, what is the major driving force moving the Earth's tectonic plates? This Section aims to find out the answer to that fundamental question.

We can distinguish two basically different types of force that might act on plates. The forces might originate essentially in the lithosphere; plates might be pushed from the ridge or pulled from the trench, or both. Alternatively, plates might be swept along by currents from the asthenosphere, responding primarily to forces originating beneath the lithosphere. We must therefore consider the forces acting on plates in two categories — those acting on the bottom surface of the lithospheric plate and those acting at plate boundaries. The forces are summarized in Table 2.3 (p.89), and are shown schematically in Figure 2.55. Refer to this diagram to help you follow the definition of plate-driving forces.

2.6.1 FORCES ACTING ON THE BOTTOM OF PLATES

We know that the lithospheric plates are in some measure decoupled from the asthenosphere because the plates move over hot spots. If the lithosphere moves over the asthenosphere, plates may be driven at least in part by asthenospheric forces. Alternatively, plates could be in part retarded by any such forces. Which happens?

The answer depends on whether the underlying asthenosphere is moving faster or slower than the plate. Let us assume initially that the asthenosphere is moving slower than the plate in the direction of plate movement. (Remember these forces are all vectors and have independent magnitude and direction.) In this case, the force acting along the bottom surface of the oceanic plate must be an **oceanic drag force**, retarding the movement of the plate. For convenience, let us label this retarding force R_{DO} (we will use R for retarding forces and F for driving forces). This force reflects resistance between the lithospheric plate and the underlying asthenosphere.

If, alternatively, the lithospheric plate is being carried along by a faster-moving asthenosphere, then R_{DO} will be a driving force (which perhaps should really be written F_{DO}), which helps the plate to move. The oceanic drag force need not be constant across the plate.

Continental lithosphere is thicker than oceanic lithosphere, so continents almost always have a 'keel' of lithospheric material. In addition, the low-velocity zone is often weakly developed or absent below the older part of continents (Block 1). Because of both these factors, the possible resistance to movement might be greater below continental plates than oceanic plates. Therefore, continental plates might be associated with an additional **continental drag force**, R_{DC}, and the resistive force acting on the base of a continental plate would be the sum of both oceanic and continental drag forces, $R_{DO} + R_{DC}$.

Remember that we cannot tell at this stage whether the lithosphere drags through the asthenosphere or whether it is propelled along by it. We can only say for certain that if the lithosphere is being retarded, the effects of the deep 'keel' beneath the continents and the relatively weak low-velocity layer will be to retard progress further. Conversely, if the plates are propelled by deep-rooted forces, the deep continental keel and its high-friction link to the asthenosphere will act rather like a sail and aid movement. In either case, oceanic drag force and continental drag force will act together in the same sense, either to help or to hinder movement.

2.6.2 FORCES ACTING AT PLATE MARGINS

Other forces that act on plates must be generated at their margins, either pushing from the ridge, acting along the sides of plates at conservative margins or dragging down at the trenches. We will investigate these by taking a close look at the ridge–plate–trench system and identifying each component force in turn.

At constructive boundaries, relatively hot, light material wells up beneath ridges creating a buoyant effect. Because of this, constructive margins typically have an ocean ridge standing some 2–3 km above the surrounding ocean floor. Here, oceanic plates must experience a force acting away from the ridge, pushing the ridge apart, simply as a result of gravity. This force is termed the **ridge-push force**, F_{RP} and acts away from the ridge (Figure 2.55). The existence of shallow earthquakes at spreading ridges shows there is in fact some frictional resistance to this force at ridges, which we can call **ridge resistance**, R_R.

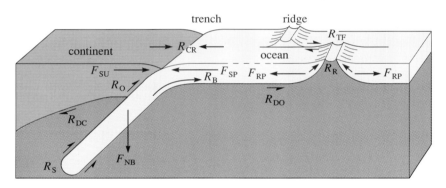

Figure 2.55 Some of the forces acting on plates.

Bounding the ridge to one side is the conservative plate margin, often represented by oceanic transform faults. At such boundaries, we would expect that plates sliding past each other would encounter resistance to movement, and earthquakes along zones like the San Andreas confirm this. We shall call this resistive force **transform fault resistance**, R_{TF}.

The situation at destructive plate boundaries is much more complex. A major component is the downward gravitational force acting on the descending slab as a result of the cold, dense slab sinking into the mantle. This gravity-generated force tries to pull the whole oceanic plate vertically down into the mantle as a result of the slab's negative buoyancy (just the opposite to the ridge). We call it the **negative buoyancy force**, F_{NB}. The plate, though, has a large surface area and cannot be pulled directly into the underlying asthenosphere, so only a component of this force is transmitted to the plate as a **slab-pull force**, F_{SP}. The slab pull is related to its gravity drive, depending on how steeply the slab is descending into the asthenosphere. The slab-pull force is greater for steeply dipping plates.

However, a slab sinking like this encounters resistance, both from drag on its upper and lower surfaces and from chemical and phase changes taking place in the slab as it accommodates to greater pressures and temperatures. We call this combined resistive force the **slab resistance**, R_S (Figure 2.55).

The downgoing plate must flex at the trench, and work must be done to achieve this. Turning energy into work acts as a further resistance, labelled **bending resistance**, R_B. There is in addition frictional resistance in the overriding plate, giving rise to shallow and deep earthquakes at subduction zones and to tectonic activity. These frictional forces can be labelled collectively as **overriding plate resistance**, R_O.

For the overriding plate, another theoretical force analogous to suction has been proposed. This **trench suction force**, F_{SU}, pulls the plate towards the trench. There are several possible causes for this force. If the subduction angle of the downgoing slab increases, the overriding plate may collapse towards the trench. The descending plate could be retreating from the overriding plate, or the descent of a cold slab might set up a convective cell, pushing the plates together. Tension could be generated by all the mechanisms that form back-arc basins (Section 2.3).

At destructive plate boundaries, we also need to define a force that reflects the collision process. This force is called the **colliding resistance force**, R_{CR}. It must act equally on both plates otherwise the trench would be constantly moving location, but of course one plate resists the push of the other. It acts in the opposite direction in each plate, but it is equal in magnitude in both of the converging plates.

A final possible force comes from the existence of hot spots. Rising hot material might contribute a force driving the overlying plates. This **hot-spot force**, F_{HS}, might be negligible where the hot spot is distant from a margin, as is the case for the Pacific hot spots. However, hot spots which are located on constructive margins, like Iceland, might exert a driving force on the surrounding lithospheric plates. We can therefore consider such hot-spot driving force as a contribution to forces pushing plates apart.

We therefore have the total possible forces acting on lithospheric plates, and these are detailed in Table 2.3.

Table 2.3   Summary of possible forces acting on lithospheric plates.

Force		Type of plate margin	Where force acts
Oceanic drag	R_{DO}		On bottom surface of plates
Continental drag	R_{CD}		On bottom surface of plates
Ridge-push and hot-spot push	F_{RP} & F_{HS}	Constructive margins	At plate margins
Transform fault resistance	R_{TF}	Conservative margins	At plate margins
Slab-pull	F_{SP}	Destructive margins	At plate margins
Slab resistance	R_S	Destructive margins	Below plate margins
Colliding resistance	R_{CR}	Destructive margins	At plate margins
Trench suction	F_{SU}	Destructive margins	At plate margins
Bending resistance	R_B	Destructive margins	At plate margins
Overriding plate resistance	R_O	Destructive margins	At plate margins

Some of these act only as driving forces, some as resistive forces, and some might act in either way.

ITQ 38

For each of the forces listed in Table 2.4, indicate their possible driving or resistive effect by ticking the appropriate space. Note that some forces might act as either a driving or a resistive force.

Table 2.4   For use with ITQ 38.

Force	Acts as a driving force	Acts as a resistive force	Might act as *either* a driving force *or* a resistive force
Oceanic drag			✓
Continental drag			✓
Ridge-push	✓		
Transform fault		✓	
Slab-pull	✓		
Slab resistance		✓	
Colliding resistance		✓	
Trench suction	✓		
Hot-spot	✓		

Present-day plate motions at present appear to be constant, so each plate must be in a state of dynamic equilibrium, controlled by a balance of driving and resistive forces. The main problem comes because it is difficult to estimate the magnitude of any single force that is acting on the plate, let alone calculate a combination of forces! But we can move forward by reasoning through the problem.

It seems unlikely that any single force is the sole driving mechanism of plate motions. For example, if the ridge-push force is the only driving force, why does the Philippine Plate, with no ridge on its boundary, move at a similar speed ($70 \, \text{mm yr}^{-1}$) to the Indian Plate, which is bounded by the Carlsberg and South Indian Ridges? Similarly, the slab-pull force cannot be the only driving force because the plates on either side of the Mid-Atlantic Ridge are moving apart without being attached to any large length of descending slab. Plate motions must be controlled by a *combination* of the forces listed in Table 2.3. We need next to consider how it is possible to work out which of these different forces contribute most to plate movement.

2.6.3   ADDING PLATE FORCES TOGETHER

We can approach the problem of discovering the most important forces that act on plates by using the reasoning process applied in the previous paragraph. For example, if the dominant driving force is ridge-push, then the fastest moving plates should be oceanic plates with the highest ratio of ridge length to surface area. Measuring ridge length and plate surface area should help solve the problem. Then we need to know which are the fastest-moving plates, so we need to know true plate motions for all plates. If, on the other hand, slab-pull is the dominant driving force, the fastest-moving plates should be the ones with the highest ratio of trench length to surface area. Also, for each kilometre of trench length, the plates with the steepest angle of subduction should be moving faster than those with lower slab angles.

Oceanic and continental drag, and slab resistance, are controlled by the true speed of plate movement. For other forces we need to know the relative motions between plates, for example to determine the colliding resistance and transform-fault resistance.

What properties of plates do we need to know to estimate all of the plate forces? Well, forces acting at plate margins should have magnitudes which are proportional to the length of ocean ridge (in the case of ridge-push), the length of ocean trench (for slab-pull, trench suction and inter-plate resistances) and the length of transform fault (for transform fault resistance). Oceanic and continental drag act over the lower surface of the plate, and so they should be proportional to the area of the plate. The magnitudes of oceanic drag, continental drag and the slab resistance are proportional to the true velocity of the plate relative to the asthenosphere. In contrast, the magnitudes of the colliding resistance and transform fault resistance depend on the relative motions of plates at plate boundaries. So, to be able to estimate the effects of each of the plate-driving forces, we must examine the rates of true plate motion in relation to the relative lengths of ocean ridge, ocean trench, transform fault, and plate area.

Table 2.5   Dimensions and true velocities of lithospheric plates.

Plate	Total plate area $\times 10^{-6}$ (km^2)	Continental area $\times 10^{-6}$ (km^2)	Average true velocity (cm year^{-1})	Circumference $\times 10^{-2}$ (km)	Length (effective length)* $\times 10^{-2}$ (km)		
					Ocean ridge	Ocean trench	Transform
(a) Eurasian	69	51	0.7	421	90 (35)	—	56
(b) N. American	60	36	1.1	388	146 (86)	12 (10)	122
(c) S. American	41	20	1.3	305	87 (71)	5 (3)	107
(d) Antarctic	59	15	1.7	356	208 (17)	—	131
(e) African	79	31	2.1	418	230 (58)	10 (9)	119
(f) Caribbean	4	—	2.4	88	—	—	44
(g) Arabian	5	4	4.2	98	30 (27)	—	36
(h) Indian	60	15	6.1	420	124 (108)	91 (83)	125
(i) Philippine	5	—	6.4	103	—	41 (30)	32
(j) Nazca	15	—	7.6	187	76 (54)	53 (52)	48
(k) Pacific	108	—	8.0	499	152 (119)	124 (113)	180
(l) Cocos	3	—	8.6	88	40 (29)	25 (25)	16

* Effective lengths (in brackets) are the lengths of plate boundary which are capable of exerting a net driving or resistive force.

Let us consider the motions of the 12 plates shown graphically in Figures 2.56–2.60 and listed in Table 2.5. Some of the important physical properties of each plate are listed together with the average true velocity of each plate (i.e. calculated relative to a hot-spot frame of reference).

For each graph, the average true velocities are plotted against total plate area (Figure 2.56), total continental area (Figure 2.57), length of ocean ridge boundary (Figure 2.58), length of ocean trench boundary with descending slab (Figure 2.59) and length of transform-fault boundary (Figure 2.60). Lengths of boundaries are plotted as percentages of the total boundary of the plate, and in Figures 2.58 and 2.59 there is a distinction between length and effective length of boundary. The effective lengths are defined as the length of boundary which is capable of exerting a net driving or resistive force. For example, two ocean ridges of equal length on opposite sides of a plate exert no net force on the plate because their effects cancel each other out.

We want now to explore whether each plate physical feature is significant in producing plate motion or not. We do this by *correlating* the data in each graph of Figures 2.56–2.60 and examining the degree and character of the correlation.

We can illustrate the method by looking at the correlations shown in Figures 2.56 and 2.57. Figure 2.56 is a plot of total plate area against true plate velocity. Let us first consider whether it shows positive or negative correlation. To show positive correlation, the plate velocity would have to increase as the plate area increased. Whilst it is true that the plate with the largest area (k, the Pacific Plate) has one of the highest velocities, it is by no means always true that plates with large areas have high velocities. The plate with the second largest area (e, the African Plate) has a relatively low velocity. The converse is not true either. The plate with the highest velocity (l, the Cocos Plate) has one of the smallest plate areas. Although it would be possible to draw a straight line through some of the data points (most notably a, b, d, e and k) which showed reasonable positive correlation (larger plates have faster velocities, and smaller plates have slower velocities), taken as a whole the data set shows no obvious pattern. It is much more like the sky at night than a line with points on it, and we can safely say that there is no correlation (or at best a very poor positive correlation) between plate area and plate true velocity. We can extend the discussion to conclude from this observation that oceanic drag is not a very significant plate-driving or plate-retarding force. If it was, we would have expected to see a good positive correlation in the former case or a good negative correlation in the latter case.

Contrast that situation with the plot shown in Figure 2.57. This is a graph of continental area on each plate with true plate velocity. Here, the plate with the largest continental area (a, the Eurasian Plate) has the lowest velocity, while the plate with the lowest continental area (l, the Cocos Plate) has the highest velocity. This is negative correlation. Whilst it is true to say that the data points in Figure 2.57 scarcely lie exactly on a line, it would be possible to draw a gently curved line through points a, b, between c and e, above d and f, between g and h but very much nearer g and through i, j, k and l. We can say that Figure 2.57 shows reasonably good negative correlation. The fact that the line is curved tells us that plates with a large continental area travel much more slowly than plates with a small continental area. It seems that continental drag is an effective resistance to plate movement — much more so than oceanic drag.

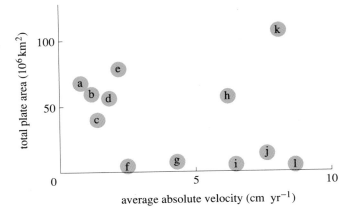

Figure 2.56   Plot of total plate area against average true velocity. The letters for the plates here and in Figures 2.57–2.60 are as in Table 2.5, i.e.: a, Eurasian, b, North American; c, South American; d, Antarctic; e, African; f, Caribbean; g, Arabian; h, Indo-Australian; i, Philippine; j, Nazca; k, Pacific; l, Cocos.

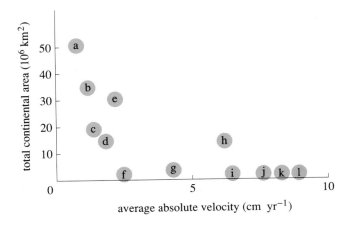

Figure 2.57   Plot of total continental area of the plates of Figure 2.56 against average true velocity.

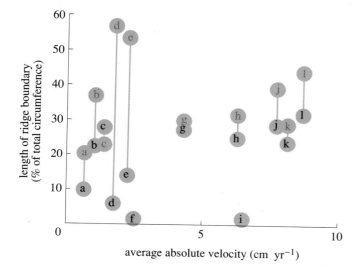

Figure 2.58   Plot of length of ocean ridge boundary (expressed as a percentage of total circumference) against average true velocity for the plates of Figure 2.56. The red letters indicate total length and the black letters indicate effective length.

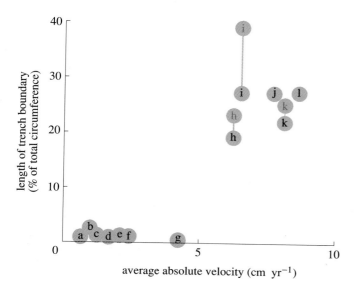

Figure 2.59   Plot of length of ocean trench boundary (expressed as a percentage of total circumference) against average true velocity for the plates of Figure 2.56. Red letters indicate total length and black letters indicate effective length.

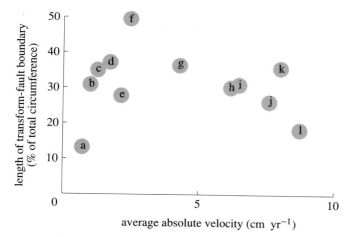

Figure 2.60   Plot of total length of transform-fault boundary for the plates of Figure 2.56 (expressed as a percentage of the total circumference of the plate) against average true velocity.

Now attempt ITQ 39 to explore the importance of each of the other factors plotted in Figures 2.58–2.60.

ITQ 39

(a)   Examine Figures 2.56–2.60 and comment on the degree and character of correlation shown by the variables (i)–(v) with average true plate velocity, by ticking the appropriate spaces in Table 2.6.

(b)   From the best positive and negative correlations that you picked out in (a), what do you conclude about the driving mechanism of plate motions?

Table 2.6   For use with ITQ 39.

	Good positive correlation	Poor correlation	Good negative correlation
(i) Total plate area (Figure 2.56)			
(ii) Total continental area (Figure 2.57)			
(iii) Effective length of ridge boundary (Figure 2.58)			
(iv) % circumference of plate attached to descending slab (Figure 2.59)			
(v) % circumference of plate that is a transform fault (Figure 2.60)			

In answering ITQ 39, you should have been able to work out that the most striking positive correlation is between true plate velocity and the length of ocean trench boundary which includes a subducted slab (Figure 2.59). The plates form two distinct data groups in this Figure. The Indian, Philippine, Nazca, Pacific and Cocos Plates are all connected to descending slabs and all are moving at 60–90 mm yr^{-1} relative to the asthenosphere. The other plates do not have significant connections with descending slabs and are moving at 0–40 mm yr^{-1}. Slab-pull is obviously an effective plate-driving force.

A less good, negative correlation is between true plate velocity and continental area (Figure 2.57), and this indicates greater drag below continents as we have already noted.

There seems to be a relatively poor correlation between ridge length and plate velocity (Figure 2.58), from which we can conclude that ridge-push is not a very effective driving force. We can combine data from several graphs to investigate this further, though. If slab-pull is effective and ridge-push isn't, is it the case that ridge push is effective in plates that have no slab? To answer this, we must select the data in Figure 2.58 that belong to plates without a slab, which we can get from Figure 2.59. This shows us we must look at plates a, b, c, d, e, f, and g (check with the Smithsonian Map to confirm these have little or no slab). These data alone show moderately good positive correlation on Figure 2.58. We can conclude from this that in the absence of a pulling slab, ridge-push is an important driving mechanism.

The best correlations from the data shown in Figures 2.56–2.60 therefore suggest that true plate velocity depends largely on the slab-pull force associated with descent of oceanic lithosphere and to a lesser extent on the ridge-push force. However, continental crust drag is a major retarding force.

SUMMARY OF SECTION 2.6

- The downward pull exerted by the cold dense slab is very large compared to other plate forces.

- The slab-pull force pulls on the attached plate and the rate of descent into the mantle increases until the force is balanced by the resisting forces acting on the descending slab.

- The observed fairly uniform rate of descent of 60–$90\,\text{mm}\,\text{yr}^{-1}$ for oceanic plates effectively represents a point of balance, the maximum velocity that can be attained by a dense slab descending into the mantle. When the forces acting downwards are balanced by resistive forces, a constant velocity is reached.

- The velocity of plates attached to the descending slabs is largely independent of characteristics such as plate area, length of ridge boundary or transform faults.

- Although slab-pull and slab resistance forces are probably the major plate-motion controlling forces, these cannot be the only forces driving plate motion because plates with little or no connection to descending slabs are also moving. Plates like these must be driven by ridge-push forces or hot spots to cause any motion at all.

OBJECTIVES FOR SECTION 2.6

When you have completed this Section, you should be able to:

2.1   Recognize and use definitions and applications of each of the terms printed in bold.

2.17 Use data relating to plate boundary lengths, plate areas, the nature of plate boundaries and spreading rates to identify the principal plate-driving forces.

Apart from Objective 2.1, to which they all relate, the two ITQs in this Section test the following Objective: ITQs 38 and 39, Objective 2.17.

You should now do the following SAQs, which test other aspects of the Objectives.

SAQS FOR SECTION 2.6

SAQ 12 (*Objectives 2.1 and 2.17*)

Summarize the evidence that distinguishes between a lithospheric or an asthenospheric origin for plate-driving forces.

SAQ 13 (*Objectives 2.1 and 2.17*)

(a)   What is the cause of the slab-pull force that drives plates?

(b)   Slab-pull is thought to be the most significant plate-driving force, but why can't it be the only one?

(c)   Which plate-driving forces seem to have little effect on the plate-driving mechanism, and why?

ITQ ANSWERS AND COMMENTS

ITQ 1

Most obviously, overlaps could mean that the plates weren't fitted together in this way! However, most overlaps are caused by features that have formed since the continents rifted apart, such as recent coral banks, recent river deltas and recent volcanoes. Overlaps don't therefore provide evidence against the fit, as it appears at first that they might.

ITQ 2

The Newcastle coalfield must have drifted to its present position from tropical latitudes either south or north of the Equator, i.e. $23°S–23°N$. This represents a latitude drift of between $23°S$ to $55°N$ ($=78°$) and $23°N$ to $55°N$ ($= 32°$). If the Earth's radius is 6370 km, its circumference must be $2\pi \times 6370$ km $= 40024$ km. $1°$ of latitude (assuming the Earth is a sphere) is therefore $40024/360 = 111$ km. The *minimum* distance Britain can have drifted since late Carboniferous times is $32 \times 111 = 3552$ km, or about 3500 km at the level of accuracy of this information. The maximum distance is at least 8658 km (say 9000 km), but could be more if Britain drifted in terms of longitude as well, or if its 'course' wasn't in a straight line.

ITQ 3

3500 km in 300 Ma represents a minimum drift rate of $3500/300 = 11.7$ km Ma^{-1}, i.e. about 12 mm yr^{-1}. The maximum drift rate could be greater than 30 mm yr^{-1}, depending on the exact drift course.

TQ 4

Let us say that two basalt samples, now lying on the same line of latitude but sited approximately 110 km apart ($1°$ of latitude), give a palaeolatitude reading differing by $1°$. The sample must have therefore rotated through $90°$. We can tell which way it rotated by investigating which sample shows the higher palaeolatitude.

TQ 5

Not strictly. The calculation that India drifted more than 5000 km is reasonable, although depending on the exact route it took the distance could be considerably more. However, India could still have subsided from high altitude to low at about $35°S$, which is hardly polar regions. It is a good example of a flawed argument based on insufficient data.

TQ 6

a)   Over the period $0–65$ Ma, India drifted from $35°S$ to about $15°N$, or some 5600 km. This is a minimum drift rate of 86 mm yr^{-1}.

b)   Over the period $65–300$ Ma, India drifted from $65°S$ to about $35°S$, some 3300 km. This is a minimum drift rate of 14 mm yr^{-1}. India's drift over the past 65 Ma might therefore be over six times the rate of the previous 235 Ma!

TQ 7

If new oceanic crust is formed at ridges, and once-joined continents move apart, the oceanic crust near to the continents should be older than the oceanic crust near the ridges. Ocean crust between these two sites might be expected to be intermediate in age between the two. Obtaining ages, either from fossil evidence or by radiometric dating, should test the theory.

ITQ 8

The shape of a magnetic anomaly (after diurnal and field-related corrections are made) depends on the shape and composition of the subsurface rock body, how deeply a strongly magnetic body is buried beneath rocks that are less magnetic, and the orientation of the magnetic field that existed when the rock body became magnetized. The anomalies must therefore represent either compositional variations in the sea-floor basalts, or systematic and highly variable sedimentary cover over a uniform magnetic body, or variations in the orientation of the Earth's magnetic field.

ITQ 9

At 75 Ma, the ocean floor has moved about 1 400 km from the ocean ridge. The rate of spreading is therefore about $19 \, \text{km Ma}^{-1}$, or $19 \, \text{mm yr}^{-1}$. As this is the rate of movement of one plate away from a ridge, we have calculated a half-spreading rate.

ITQ 10

Even if the age of oceanic crust being destroyed can be identified using its linear magnetic anomalies and the magnetic polarity time-scale, there is no simple means of estimating the time when the destructive boundary was initiated, and hence the rate of movement since this time.

ITQ 11

Using the appropriate distance scale on the Smithsonian Map which allows for the distortion of the map's projection at 50° S, the 4 000 m depth contour lies about 800 km from the spreading ridge in the Pacific Plate. From Figure 2.11, this oceanic crust must be about 20 Ma old. So, 800 km of movement in 20 Ma represents a relative movement of $40 \, \text{mm yr}^{-1}$. The equivalent depth contour in the Antarctic Plate is about 900 km from the ridge, indicating a relative movement of $45 \, \text{mm yr}^{-1}$. Note that both these figures are quite different from the ones actually written on the Smithsonian Map. This is because (among other things) our calculations give relative rates, whereas we are told the Smithsonian Map shows absolute spreading rates.

ITQ 12

The completed Table is shown below and explained further in the text.

	(i) Rate of plate motion	(ii) Direction of plate motion
(a) American Plate fixed	$40 \, \text{mm yr}^{-1}$	East
(b) Australian Plate fixed	$40 \, \text{mm yr}^{-1}$	West

ITQ 13

The age range along the Hawaiian chain is 42 Ma. This is spread over a distance of some 30° of longitude, or about 3 000 km. This gives a calculated rate of movement of some $70 \, \text{km Ma}^{-1}$ (equivalent to $70 \, \text{mm yr}^{-1}$). From the Smithsonian Map, we can see that the Hawaiian chain lies approximately WNW of the current hot-spot volcanoes. The Pacific Plate has therefore been moving west-northwestwards over the hot spot at about $70 \, \text{mm yr}^{-1}$ for the past 42 Ma (Note that this is similar to, but not exactly the same as, the arrow marked on the Smithsonian Map.)

ITQ 14

In the Atlantic, stripes are roughly symmetrical about a central ocean ridge, which is also where the youngest oceanic crust is to be found. Oceanic crust of similar ages is found on both continental margins. In the Pacific, the youngest oceanic crust is found near a ridge that is well into the eastern part of the ocean, far from being centrally sited. Oceanic crust is also much younger at the South American continental margin than it is near Japan and the Philippines. Crust older than 80 Ma is no longer preserved in the eastern Pacific. We can also note that oceanic crust is not the same age along the western South American coastline, whereas along the eastern South American coastline it is.

ITQ 15

Either (i) South America might not have split away from whichever continent it was joined to until 65 Ma ago (yellow stripe), or (ii) oceanic crust formed longer ago than 65 Ma has somehow been removed from this continental margin. The existence of oceanic crust of a range of ages next to the South American continental margin supports the latter reason.

ITQ 16

The most obvious candidate is a less dense rock type; perhaps a sedimentary rock or granite. However, sediments and granites are not major constituents of the mantle. If there is no compositional change over the Bouguer anomaly, there must be a physical change in 'normal' mantle that makes it less dense.

ITQ 17

Figure 2.18 shows that oceanic crust is strongly layered down to a depth of 10–15 km below sea-level. Both seismic velocities and observed rock compositions suggest that the layers of oceanic crust consist of pillow basalt or basalt-like rocks overlain by deep-ocean sediment. Below the basalts, there is a complex of multiply-intruded dykes. Beneath this, in turn, is a substantial thickness of mostly gabbro (the coarse-grained equivalent of basalt).

ITQ 18

In the physical sciences, data are often generated by experiment and then analysed mathematically. In some areas of technology like rocket design, a working scale model might be built and tested. In architecture and civil engineering, the effective testing of some aspects of a building may have to wait until the building itself is built. The failure of structures during earthquakes shows that effective advance testing may not always be possible. In other learning areas, actual analogue models are often not possible or practical. Students trying to probe the battle strategy of William the Conqueror in 1066 have to rely on 'thought models' gleaned mostly from accounts written at the time or shortly afterwards.

ITQ 19

At this stage, we can rule out experimentation and mathematical analysis. It is possible, but not straightforward, to duplicate the pressure, temperature and other physical conditions that we know exist at spreading ridges, but we can hardly model either the physical size or the time dimension. Analogue models are therefore difficult. We fall back on empirical or 'thought models' under these circumstances. There are surprisingly close links between the thought processes in Arts and Science.

ITQ 20

Plate 2.2 shows clear variations in shape and form of the East Pacific Rise. The ridge in the upper part of the view seems to die out towards the lower centre, to be replaced by a parallel ridge to the west. There are significant topographic variations where the ridges overlap, too.

ITQ 21

In Figure 2.61a, the geological feature has been displaced leftwards by the fault. The fault has a sinistral displacement. It has the same displacement whether you look at it the right way up or turn the page upside down (i.e. whichever side of the fault you are standing on). In Figure 2.61b, the plates are moving apart at the ridges, so the two plates are moving rightwards (dextrally) past each other at the transform fault. Again, it doesn't matter which side of the fault you are standing on, the displacement remains the same. The sense of displacement on a transform fault is exactly the opposite of what you might expect, based on faults that displace 'ordinary' geological features.

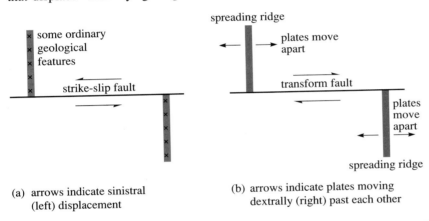

(a) arrows indicate sinistral (left) displacement

(b) arrows indicate plates moving dextrally (right) past each other

Figure 2.61   For use with answer to ITQ 21.

ITQ 22

(a)   The relief of the transform zone is both greater and more irregular than the two segments of East Pacific Rise we can see. From the colour scale bar, there appears to be about 200 m of relief along the transform.

(b)   There is a clear zone of irregular relief east of the north segment of the East Pacific Rise. This most probably locates a fracture zone.

(c)   The northern plate is moving westwards past the transform while the southern plate is moving eastwards. (Both plates must be moving away from their spreading ridges.) This is therefore a sinistral (left-lateral) transform. Note though that the East Pacific Rise *appears* to be displaced rightwards, or dextrally.

(d)   The two features labelled 'RTI' are the junctions between the spreading ridge and the transform; the ridge–transform intersections.

ITQ 23

On its eastern and southern margins, the Pacific Plate is moving away from its neighbours. Along the western seaboard of North America, the Pacific Plate has a complex movement pattern, but broadly is moving parallel to the boundary of the North American Plate. A similar situation exists north of Australasia. The Pacific Plate is converging orthogonally (movement direction at right angles to the plate boundary) in the following areas (north to south):

- The northeastern part of the Aleutian arc, near Alaska.

- East of Kamchatka, the Kuril Islands and Japan.

- East of the Mindanao Islands.

- In parts of the South China Sea.

- East of Tonga and south beyond New Zealand.

(Note that the Smithsonian Map includes rather more areas in the diagram in the explanatory notes.)

Each of these areas is a zone of intense tectonic, seismic and volcanic activity.

ITQ 24

They all are!

ITQ 25

As earthquakes with foci at about 100 km depth lie about 100 km from the trench, and earthquakes with foci about 600 km down lie about 600 km from the trench, the failure plane must have a slope of about 1 in 1, i.e. about 45°.

ITQ 26

No, this angle is not constant for the whole Tongan Trench. Shallower earthquakes lie further away from the trench between Fiji and Samoa, indicating that the dip of the failure plane is less than 45° here. In fact, it is nearer 30°, but still dips westwards.

ITQ 27

The shape of the present-day heat flow curve is quite unlike those predicted in the early stages of subduction, and shows higher values than might be predicted for the first 50 Ma of subduction. The curve looks most like the model heat flows for 80–100 Ma after initiation. This implies that subduction commenced there at least 80 Ma ago

ITQ 28

The ocean floor at present entering the Japan Trench is about 120 Ma old. If the subduction zone has been active for over 80 Ma, assuming similar convergence rates to the present day situation, the plate itself must be at least 200 Ma old. It might of course be much older if convergence was faster over the past 80 Ma, or somewhat younger if convergence was slower then.

ITQ 29

The answer lies in the closeness of the South American continent. This part of the Atlantic is in the distributary system of the Orinoco River. The Orinoco drains much of equatorial northern South America and transports a vast amount of sediment. This is deposited into the Atlantic near the south end of the Lesser Antilles subduction zone and washes out onto the Atlantic Plate as sediment known as turbidites.

ITQ 30

Oceanic crust is forming at spreading axes as a result of extension, so a tensional rather than a compressional regime is suggested by oceanic crust in a back-arc basin. This is rather surprising, since convergent plates would indicate compression rather than tension.

ITQ 31

Since the magnetic anomalies associated with an RRR junction would be formed in the Y-shaped pattern shown in Figure 2.35a, an obvious way of identifying such former triple junctions would be to look for magnetic anomaly patterns with this shape.

ITQ 32

(a)   The relative motion of the Pacific Plate is consistently from right to left in Figure 2.40. In geographic terms, this is towards the northwest.

(b)   In passing from southeast to northwest (from right to left), the American margin of the plate changes from constructive through conservative to destructive.

ITQ 33

The accurate answer to this depends on the subduction rate relative to the spreading rate. Given the present balance, in time the East Pacific Rise will eventually reach the Aleutian–Kurile–Japan destructive margin and that part of the oceanic crust will be subducted. However, note that in the past it has migrated eastwards!

ITQ 34

The tip has propagated westwards about 200 km in 7 Ma. This represents a propagation rate of about $30\,\text{mm yr}^{-1}$.

ITQ 35

Geoscientists would need to know whether one or other of the adjacent continental blocks had moved from their present positions over time. This could be done by taking palaeomagnetic samples from suitable rocks of different ages on each continent, to test for apparent polar wander and hence continental drift. Alternatively, it could be approached by investigating the magnetic stripe pattern of adjacent oceans, to estimate relative continental configurations over the geologically recent past.

ITQ 36

The southern tip of India has drifted to its present position at $10°\,\text{N}$ from a position 115 Ma ago of about $45°\,\text{S}$. This represents a minimum distance of $55°$ of latitude or 55×111 km (the figure we calculated for $1°$ of latitude in ITQ 2). This is a minimum distance of 6 100 km, and a drift rate over that period of $6\,100 / 115 = 53\,\text{km Ma}^{-1}$, or about $50\,\text{mm yr}^{-1}$.

ITQ 37

The third type of oceanic plate boundary not discussed so far is a transform plate boundary. It is characterized by lateral displacements, rather than plate movements towards or away from the boundary.

ITQ 38

Table 2.7 Answer to ITQ 38.

Force	Acts as a driving force	Acts as a resistive force	Might act as *either* a driving force or a resistive force
Oceanic drag			✓
Continental drag			✓
Ridge-push	✓		
Transform fault		✓	
Slab-pull	✓		
Slab resistance		✓	
Colliding resistance		✓	
Trench suction	✓		
Hot-spot	✓		

ITQ 39

(a)   A completed list of correlations of absolute velocity with variables is shown below (Table 2.8). If you disagree with any of the answers, you should look again at Figures 2.56–2.60.

Table 2.8 For use with answer to ITQ 39.

	Good positive correlation	Poor correlation	Good negative correlation
Total plate area (Fig. 2.56)		✓	
Total continental area (Fig. 2.57)			✓
Effective length of ridge boundary (Figure 2.58)		✓	
% of plate circumference attached to descending slab (Fig. 2.59)	✓		
% of plate circumference that is a transform fault (Fig. 2.60)		✓	

(b)   The good positive correlation of absolute plate velocity with the percentage of circumference attached to a descending slab implies that the slab-pull force (F_{SP}) is a major driving force of plates.

SAQ ANSWERS AND COMMENTS

SAQ 1

(a)   From astronomical techniques like VLBI and GPS, we know whether the continents are actually drifting apart at the present time. Use of these techniques has confirmed continental drift. For example, Europe and America are presently moving apart at around $20 \, mm \, yr^{-1}$.

(b)   Geoscientists can locate and date magnetic reversal stripes over the Earth's recent past. Volcanic islands that have developed over hot spots, like Hawaii, can be dated and show that oceanic plates move relative to the global constellation of hot spots. Earthquakes are in fact generated by movements between blocks of the Earth's crust, and they are commonly found along plate boundaries, suggesting constant plate movement. Geological features which once were continuous, like the cratonic areas and mobile belts of South America and Africa, are now separated by oceanic material. This suggests that the continents have moved apart for the time-span indicated by the age of the oceanic lithosphere.

SAQ 2

Basalts are erupted onto the ocean floor at oceanic spreading centres. Basalts contain relatively high amounts of the magnetic iron oxide magnetite, which acquires a magnetization in the same direction as the Earth's ambient magnetic field. Since the Earth's magnetic polarity switches from north to south from time to time, ocean-floor basalts acquire a striped magnetic pattern which relates intimately to their age. Geoscientists study these stripes to erect a magnetic time-scale, that is a chart of magnetic reversals over the Earth's recent past. Comparing the record of stripes in any particular part of ocean floor with the standard magnetic time-scale can give the age of that part of ocean floor and hence the half-spreading rate at the relevant ocean ridge. Magnetic stripes also provide a remote means for dating ocean basalts, and for palaeolatitude calculations.

SAQ 3

Evidence from volcanic island chains like Hawaii in the Pacific and Tristan da Cunha in the Atlantic tells us that plates move over hot spots. Comparing the observed tracks of hot-spot trails with those predicted from the movement of adjacent plates tells us that all the known hot spots are fixed relative to each other. Comparing palaeolatitudes given by the hot-spot frame over time with their predicted positions (from polar wandering curves) confirms that the hot-spot frame and the Earth's core are moving relative to each other, but only slightly. Whilst the hot-spot frame is thus known not necessarily to be stationary, its movement over time (if any) is slight. The hot-spot reference frame therefore provides the best possible evidence for the actual movement of plates with time.

SAQ 4

Earthquakes occur at every type of plate boundary, so their location can't be used to distinguish between plate boundary types. Shallow-focus earthquakes (<60 km deep) could be associated either with an ocean ridge system, with transform faults or with the oceanward side of a Wadati–Benioff zone. Deep-focus earthquakes (>60 km deep) are almost exclusively associated with destructive plate margins, either island arcs or ocean–continent destructive margins. Strong earthquakes typically occur at destructive margins, while within-plate areas are characteristically quiet, seismically. Therefore, distributed zones of shallow and deep earthquakes, some of them strong, characterize destructive plate margins, while exclusively shallow linear earthquake zones characterize both conservative and constructive plate margins. Earthquake data alone are not sufficient to tell these latter two apart.

SAQ 5

Table 2.8   Answer to SAQ 5.

Geoscientific feature	Destructive margin	Constructive martin	Conservative margin
Weaker earthquakes (magnitude <7.5)	✓	✓	✓
Stronger earthquakes (magnitude >7.5)	✓	×	×
Shallow-focus earthquakes (<60 km)	✓	✓	✓
Deep-focus earthquakes (>60 km)	✓	×	×
Active volcanoes	✓	✓	×
Wadati–Benioff zones	✓	×	×
Transform faults	×	×	✓
Offset magnetic stripes	×	×	✓
Missing magnetic stripes	✓	×	×
Higher heat flow	×	✓	×
Lower heat flow	✓	×	×
Median rifts	×	✓	×
Magma generation	✓	✓	×
Accretionary prisms	✓	×	×
Fracture zones	×	×	✓
Island arcs	✓	×	×

SAQ 6

The stability of plate triple junctions largely depends on the relative movements of the three component plates. Relative movement of plates away from the triple junction is stable and generates an RRR, RRF or RFF type of junction. Relative movement of plates towards the triple junction is basically unstable (unless certain special geometric conditions are met) and produces TTT, TTF and TFF junctions. Any triple junction involving trenches will change constantly as one or more plates become subducted. FFF junctions are also unstable and must evolve immediately to either an RFF or an RRF junction.

Summarizing, movement towards the triple junction (one or more trenches) promotes instability, whereas movement away from the triple junction (ridges) promotes stability.

SAQ 7

The Farallon Plate is being subducted beneath the North American continent faster than new oceanic crust is being generated at its ridge. The ridge system is therefore moving towards the North American continent. Coupled to this, the northern RTT and the southern FFT triple junctions between the Pacific, Farallon and North American Plates are moving towards each other. In time, therefore, the Farallon Plate will be wholly subducted beneath North America and the whole western seaboard of North America will become a subduction–transform complex.

SAQ 8

This is a tough question! Future plate tectonic configurations depend on the absolute motion of the Pacific Plate and of adjoining plates relative to it. The absolute motion of the Pacific Plate is 80–100 mm yr^{-1} (depending on the exact location of the vector on the Earth's spherical surface) towards the WNW. This means that the plate will travel 80–100 km Ma^{-1} or between 4 000–5 000 km in the next 50 Ma, assuming constant rates of movement. In turn, this means that all of the Pacific Plate between Tonga and Baja California will have been subducted beneath the Aleutian–Japan–Philippine arc system, including the hot-spot ridge of Hawaii. Island arc volcanism is likely to continue in the western Pacific.

In the NE Pacific, much of California will be displaced northwestwards, to accrete onto western Alaska. Parts of the SW Pacific may accrete as microcontinents onto the subduction margins of the western Pacific.

Depending on the rate of subduction versus plate creation in the eastern Pacific, the Pacific Plate could grow to cover the entire Pacific Ocean floor (subduction > creation), leaving almost all the American margin as a dextral transform zone. Alternatively, the Pacific Plate could shrink (subduction < creation) and the Nazca–Cocos Plate system expand accordingly.

SAQ 9

The plate tectonic cycle for oceanic crust begins with the formation of new lithosphere at ocean ridges which is then transported to subduction zones where it is recycled into the mantle. By contrast, continental crust is thinned and splits at spreading centres, and is then passively transported towards an opposing continental margin. On collision, two continental fragments accrete together to form a new block of continental crust.

The cycles are so distinctly different because:

(i) dense oceanic crust is subducted whereas lighter continental crust is not;

(ii) oceanic crust is formed at spreading centres whereas continental crust thins, stretches and splits;

(iii) continental crustal blocks accrete together at destructive margins whereas oceanic crust is subducted and destroyed.

SAQ 10

The East African Rift valley might be expected to undergo a similar history to that recorded by the Gulf of Aden or the North Sea. Crustal stretching has generated normal faults that bound the rift valley. Basaltic magmas have been generated by stretching-related melting of the mantle. If rifting continues, eventually wholly basaltic material will floor the rift and the Horn of Africa will drift eastwards relative to continental Africa, becoming a microcontinent.

SAQ 11

The roots of an old continental collision zone should be most easily recognized by the presence of folded and thrust metamorphic rocks. If an Andean margin had been established prior to collision, extensive granite plutons should be present. The whole structure would have been underlain by a deep crustal root, but that may be modified or reduced if the region has adjusted isostatically since collision. Geoscientists might discover an ophiolite which would represent part of non-destroyed oceanic crust which once separated two continents. Granite bodies dated at the age of collision might also be expected. Interestingly, all these features can be seen in the Scottish Highlands, dating from continental accretion 600–400 Ma ago.

SAQ 12

The most significant single piece of geoscientific evidence for a lithospheric 'drive' for plate tectonics comes from hot-spot tracks. Hot spots originate in the asthenosphere and all the lithospheric plates that make the present configuration track over the array of hot spots. Plate movement must therefore be confined to the crust and upper mantle. Deep or whole mantle forces would move the hot spots, too.

Furthermore, the strong links between subduction zones and plate velocity and ridge length and plate velocity both suggest that gravitational forces acting on the plate margins are the major driving forces.

SAQ 13

(a) Slab-pull is caused by gravity. The cold, dense subducted plate sinks under its own weight into a hotter mantle, until it is completely dewatered and of a comparable density and temperature to its surrounds. Slab-pull is dependent on slab descent angle, and the slab-pull force is greater for steeply dipping subduction zones.

(b) Slab-pull cannot be the only plate-driving force, because at least one moving plate, the African Plate, has no attached subducting oceanic plate. Whilst the rate of movement of the African Plate is slow, it is actually moving. Similarly, the South American Plate is moving and has no attached subducting slab, although it is linked to a subduction zone on its west side.

(c) There is little correlation between total plate area and plate velocity, between transform faults and plate velocity or between overriding plate thickness and plate velocity. Therefore, transform fault resistance, mantle drag and friction at subduction zones seem to play little part in driving or retarding moving plates.

SUGGESTIONS FOR FURTHER READING

KEAREY, P. and VINE, F. J. (1990) *Global Tectonics*, Blackwell Scientific Publications, Oxford, 302pp.

A good general text about all aspects of plate tectonics, although it goes into considerably greater depth than required for this Course. Useful parallel reading if you have the time. You'll recognize many of the diagrams!

PARK, R. G. (1988) *Geological Structures and Moving Plates*, Blackie & Sons, Glasgow, 337pp.

A specialized text dealing mostly with continental rock deformation and its relationship with plate tectonics. Significantly higher level than this Course, but a very useful compendium of structural geology for those interested in this aspect. Some parts much more readable than others.

ACKNOWLEDGEMENTS

Very many thanks to colleagues on the Course Team for improving early drafts even before they were written! Special thanks to them, to David Darbishire and Colin Hayes, and to Ben Kneller and Ray Macdonald for commenting on and improving the working draft.

Grateful acknowledgement is also made to the following sources for permission to reproduce material in this Block:

Figure 2.2 reprinted with permission from *Nature*, **225**, pp. 139–144, copyright © 1970 Macmillan Magazines Ltd; *Figure 2.3a* A. L. Du Toit (1972) *Our Wandering Continents,* Oliver & Boyd; *Figure 2.3b Nature,* **238**, pp. 92–3, copyright © 1972 Macmillan; *Figure 2.3c* C. McA. Powell *et al.* (1980) 'A revised fit of east and west Gondwanaland', *Tectonophysics*, **63**, Elsevier; *Figure 2.4* A. Hallam (1975) 'Alfred Wegener and the hypothesis of continental drift', copyright © 1975 Scientific American Inc, all rights reserved; *Figure 2.5* P. Kearey and F. J. Vine (1990) *Global Tectonics,* Blackwell; *Figure 2.6* K. M. Creer (1965) 'Palaeomagnetic data from Gondwanic continents', in Blackett, P. M. S. *et al.* (eds) *A Symposium on Continental Drift,* The Royal Society; *Figure 2.7* H. W. Menard (1964) *Marine Geology of the Pacific,* McGraw-Hill; Figure 2.8 M. H. P. Bott (1982) *The Interior Of The Earth, Its Structure, Constitution and Evolution,* 2nd edn, Edward Arnold; *Figure 2.13* G. B. Dalrymple *et al.* (1973) 'The origin of the Hawaiian islands', *American Scientist*, **61**, p. 305, © 1973 Scientific Research Society (Sigma XI); *Figure 2.15* R. A. Duncan and M. A. Richards (1991) 'Hot spots, mantle plumes...', *Review of Geophysics*, **29**, No. 1, © American Geophysical Union; *Figure 2.16* V. Courtillot and J. Besse (1987) 'Magnetic field reversals...', *Science*, **237**, p. 1142, © AAAS; *Figure 2.17 Annual Review of Earth and Planetary Sciences*, **10**, © 1982 Annual Reviews Inc; *Figures 2.21a,b* and *2.22 Nature*, **302**, pp. 56–7, copyright © 1983 Macmillan; *Figure 2.22 Nature*, **302**, p. 56, copyright © 1983 Macmillan; *Figure 2.23 Nature*, **335**, p. 224, copyright © 1988 Macmillan; *Figure 2.26* L. R. Sykes 'The seismicity and deep structure of island arcs', *Journal of Geophysical Research*, **71**, No. 12, p. 2987, © American Geophysical Union; *Figure 2.27* A. Hasegawa *et al.* (1978) 'Double planed deep seismic zone...', *Geophysical Journal of the Royal Astronomical Society*, **54**, p. 288, © 1978 Royal Astronomical Society; *Figure 2.28* D. E. Hayes and M. Ewing (1971) 'Pacific boundary structure', in A. E. Maxwell (ed.), *The Sea*, Vol. 4, Pt 2, copyright © 1971 Wiley; *Figure 2.31a* R. G. Park (1988) *Geological Structures and Moving Plates,* Blackie; *Figure 2.31b* J. Casey-Moore (1982) 'Offscraping and underthrusting...', *Bulletin of the Geological Society of America*, **93**, p. 1066, © Geological Society of America; *Figure 2.32* G. K. Westbrook (1982) 'The Barbados ridge complex', *Special Publ. Geol. Soc. Lond.,* Blackwell, © 1982 The Geological Society; *Figure 2.33* D. E. Karig (1974) 'Evolution of arc systems in Western Pacific', in Donath, F.A. *et al.* (eds), *Annual Review of Earth Planetary Sciences*, © 1974 Annual Reviews Inc; *Figure 2.40 Nature*, **216**, p. 1278, © 1967 Macmillan; *Figure 2.41 (left)* A. Cox and R. B. Hart (1986) *Plate Tectonics: How It Works,* Blackwell; *Figure 2.41 (right)* C. H. Scholz *et al.*, 'Late Cenozoic evolution of the Great Basin...', *Bulletin of the Geological Society of America*, **82**, © Geological Society of America; *Figure 2.43* P. A. Ziegler (1982) *Geological Atlas of Western and Central Europe,* Shell; *Figures 2.44* and *2.53* illustrations by Ian Worpole (1986), in Molnar, P. (1986) 'The structure of mountain ranges', *Scientific American*, July; *Figure 2.49* illustration by Todd Pink (1983), in Clark Burchfield, B. 'The continental crust', *Scientific American*, Sept; *Figure 2.52 Nature*, **311**, p. 616, copyright © 1984 Macmillan; *Figure 2.55* M. H. Bott (1982) *The Interior of the Earth: Its Structure, Constitution and Evolution,* Edward Arnold.

INDEX FOR BLOCK 2

(**bold** entries are to key terms; *italic* entries are to tables and figures)

absolute plate motion **19**
  see also true plate motion
accretion **54**, 76, **83**
  of California to Alaska 107
  of continental blocks 107
accretionary prisms **54**, 56, 61, 62
  compression of 76–7
  thickened by stacking 54
accretionary wedge, folds and faults in 76, *76*
active margins **29**, 46
Aden, Gulf of 107
  a divergent plate margin 73–5
  separation reconstruction *75*
Aegean region 84
Afar depression 74
Africa–S. America, geological fit 9–10, *10*, 29
African Plate 20, 25, 91, 108
  major plate boundaries and tectonic features *73*
Aleutian Islands 47, 57
Aleutian Trench, bent magnetic striping *59*, 59–60
Aleutian trench–arc system *51*, 51–2, 55
Alpine Fault, New Zealand 57–8, *58*, 84
Andean margins **77**
Andes 46, 77–8
Antarctic Plate 23, 25, 98
apparent polar wander 12–13, 102
apparent polar wander path 26, 29
apparent polar wandering curve **12**, 13, 105
Aqaba, Gulf of 83
Arabia, movement of 83
Arabian microplate 84
arc migration 55
arc–continent collisions **83**
aseismic ridges 38
asthenosphere, rising *39*, 70, 72
asthenospheric forces, acting on plates 87, 108
asymmetric spreading **63**
Atlantic Ocean 28, 29, 99
  continental fit around *7*, 7–8, 97
Atlantic Plate 53
Austral–Gilbert–Marshall Islands seamount chain *see* Marshall–Ellice Islands–Austral seamount chain
Australian–Antarctic Rise, half–spreading rates at 63
average rate of movement **18**
Aves Ridge *53*, 55

back-arc basins *53*, **56**, 61
  generation of *56*, 57
Baja California peninsula 73, 83, 84
Barbados Ridge *53*, 54, 56
basalt 30, 44, 60, 105
  at ocean ridges *see* pillow lavas

magnetite crystals in 11–12
ocean-floor, and magnetic stripes 14–16, 30, 31
palaeolatitude recorder 12
producing local magnetic anomaly 14
basement, overthrust *78*
Basin and Range Province, USA 83–4
bending resistance **89**
block rotation *72*, 72
Bouguer anomalies 37
  Mid–Atlantic Ridge *35*, 35
  negative 54
Bouvet Island hot-spot trail 25
Britain, drift of since coal formation 10, 97
brittle failure/fracture 70, 71, *72*, 84
bulges, oceanwards of trenches 49, 60
  caused by flexure of rigid lithosphere 50
  positive anomalies over *51*, 51–2

Canary Islands hot-spot trail 25
Canary Islands 20
Cape Verde Islands hot spot trail 25
Caribbean Plate *53*, 53
Carlsberg Ridge 90
  relative spreading rate for 19–20, *20*
climatic indicators, of continental drift 10–11, 29
Clipperton fracture zone *41*
Clipperton Transform fault 43–4, 45
  case study 45–6
coal, climate of deposition shows continental drift 10, 97
Cocos Plate 23, 43, *53*, 91
  propagating ridge segment *64*, 64
Cocos Ridge *53*
colliding resistance force **89**, 103
combined plate boundaries **33**, **57**, 57–8
compression 46, 54–5, **76**, 77, *78*, 84
confining pressure, effect of reduction in 38
conservative margins 88, 106
  within continental crust 82–3, 85
conservative plate boundaries **17**, 18, **33**, **45**, 60, 61
conservative plate boundary fault 66
constructive margins 23, 25, 36
  within continental crust 70–5, 84, 106
constructive plate boundaries **16**, 17, 30, **33**, 33–42, *43*, 60, 88
  see also mid–ocean ridges; spreading axes; spreading ridges
continental breakup 84
  timing of 28–9
continental collision 78–82, 86, 107–8
  case study: the Himalayas 79–82
  oblique 83, 85
continental crust
  age of 69, 70

average densities, upper and lower crust 69, 70
conservative margins within 82–3
constructive margins within 70–5
differing from oceanic crust 69–70, 84
increased at oceanic subduction zones 75–6
less uniform 70–1
modifying plate boundaries 77
see also continental collision
relocation of 76
silica-rich 69
thickened and deformed *78*, 78
too light for subduction 81, 84
continental drag force **87**, 90–1, 94, 103
  effective resistance force 92
continental drift 7, 10–11, 29, 102
continental plates **69**
  boundaries as broad zones 70–1
continental reconstruction, geometric 7–8
continental transforms 82–3
  linking subduction zones 84
continents
  age of 84
  drifting apart 104
  geological match between 8–11, 29
  keel of lithospheric material 87
  see also Africa; Indian continent; South America
convergent margins 49, 56
  as destructive plate boundaries 52
convergent plate boundaries *see* destructive plate boundaries
crustal roots **77**, 107
crustal shortening 76–7, *78*
crustal stretching 38, 71, 75, 86, 107
crustal thickening 84
  at island arcs 55
  beneath the Himalayas 81–2, *82*
  upper crust 77

Dead Sea fault zone 83, 85
deformation *76*
  brittle 70, 71, *72*, 84
  rock deformation **54**
  see also ductile processes
destructive margins 102, 105
  at the ocean–continent boundary 75–8, 84, 105
  and continental collision 78–82
destructive plate boundaries **16–17**, 18, **33**, 46–57, 60, 88, 98
Dietz, R.S. 13
divergent boundaries 4, 60
  failed 72
divergent plate margins
  within continents 71–3
  case study: Gulf of Aden 73–5
downgoing slab 51, 60, 88, 89
  roll–back of *56*, 57
  thermal effects of 52
  and volcano formation 55

ductile processes *72, 76*, 77, 84
dykes *see* sheeted dyke complexes

earthquake epicentres 4
earthquake foci 47, 60
    round Tonga Trench *48, 48*, 49
earthquakes 4
    Andes 78
    associated with movement of rock
        bodies 47, 105
    caused by downgoing slabs 51
    deep 46, 47, 62, 105
    and failure planes 48–9
    from internal deformation 51
    shallow 46, 47, 62, 88, 101, 106
    at mid-ocean ridges 33, 34, 37–8, 43,
        105
    oceanward of deep landward
        earthquakes 47, 105
    weaker and stronger 62
East African Rift, crustal stretching 86,
107
East Pacific Rise 16, *34*, 43, 45, 100
    case study 40–2
    and San Andreas Fault zone 83
    subduction of 66, *67*, 102
Easter Island 22
exotic terranes **83**
extension 57, **70**, 70–5, *71*, 84, 101
    associated with convergent margins
        83–4
    by opposed flow 57

failure planes 101
    associated with trenches 49
    Japanese *49*, 49
    Tonga–Fiji 48–9
Farallon Plate, subduction of 66, *67*, 68,
106
fault zones 82
faults 71, *76*, 76, 84
    active, spreading ridges *34*, 35
    normal *72*, 78, 84
    strike–slip 100
fissure zone, spreading ridges *34*
fixed frame of reference **20**
    see also hot–spot reference frame;
        magnetic reference frame
flood basalts, Bangladesh 25
folds **54**, *76*, 76, 77, 84, 107
fore-arc basins *53*
fore-arc ridges *53*
fracture zones *41*, **45**, 46, 62, 100
free-air anomalies
    Aleutian trench–arc system *51*, 51
    Mid-Atlantic Ridge *35*, 35
    negative 54
frictional resistance, of overriding plate
89, 108

gabbro 39, 60, 99
    layered 37, *39*
geological record, and continental drift
10–11
glacial deposits 10

showing drift of Southern continents
    11
global positioning by satellite (GPS) 28,
105
GLORIA sonographs *40*, 40
Gondwanaland **79**
graben **71**, 84
granodiorite 69
gravitational force 88, 108
Great Magnetic Bight *59*, 59–60
Grenada Trough *53*, 56

half-spreading rate **17**, 18, 63, 98
    relative 19
Hawaii 20, 22, 105
Hawaiian hot spot 22
Hawaiian Ridge 4, 16, 22, 107
Hawaiian Ridge–Emperor seamount
chain *21*, 22, 98
heat, radiogenic 69
heat flow 4, 62, 84
    age of Japanese subduction zone 52,
        101
    and a descending cold slab 52, 60
    high 4, 47
    at spreading ridges 34, 37, 38
    variations across trench–arc systems
        52
heat flow low 47
Himalayas
    Andean-type processes and
        continental collision 80–2, *81*
    and continental collision 79–82
Horn of Africa 107
hot spots 4, **21**, 89, 95
    changing plates 25
    and formation of oceanic crust 38
    prediction of on plates adjacent to
        Pacific Plate 23
    stationary or not 22–5, 25–7
hot-spot chains, parallel nature of 25
hot-spot force **89**, 103
hot-spot reference frame 20–2, 30, 31,
105
hot-spot trails 25, 105, 108
    modelled *24*

Iceland 8
    effect of hot spots 89
Indian continent
    drift of 11, 12, 79–80, 97, 102
    too buoyant for subduction 81
Indian Ocean 28
Indian Plate 90
    speed of movement 25
    subduction beneath Tibet *81*, 81
Indo-Australian Plate 25, 49
instantaneous rate of movement **18**
intrusions, granite *see* plutons
island arcs **46**, **47**, **55**, 62
isostatic imbalance, due to flexural
rigidity 51–2

Japan Trench 101

Japanese islands 84
Japanese subduction zone 52
Juan de Fuca Ridge *14*, 45

Kerguelen hot spot 25
Kodiak Island–Cobb seamount chain *21*,
22
Kohistan island arc 83

lava 60, 76
leaky transforms **57**, 61
Lesser Antilles subduction zone 52–7,
75–6, 101
lithospheric plates *see* plates
lithospheric thinning 71, 84
Lord Howe Ridge 84

Macdonald seamount 22
magma
    at ocean ridges 36–7
    basaltic 107
    generation of 55, 62, 77, *78*, 84
magma chambers 37, 38, *53*, 60
    case study: East Pacific Rise 40–2
    eruption through vertical fissures 38
    fed by pulses of molten material 42,
        43
    as open systems 37
    shallow 43
    size of 39
    see also ridge segments
magnetic anomalies **14**, 98
    change in orientation of 63–4
    ocean-floor pattern of *14*, 15, 102
magnetic axis, possible wander of 26
magnetic field 11, 14, 15
    and the spin axis 27
magnetic inclination **11**
magnetic minerals, orientation of 26
magnetic poles, fossil *see* palaeopoles
magnetic reference frame, and true polar
wander 25–7
magnetic reversals *14*, 15–16, 105
    for estimating average sea-floor
        spreading rate 18
magnetic stripes 14–16, 26, 30, 31, 62,
63, 102, 105
    and age and preservation of oceanic
        crust 29, 99
    bent/asymmetric *59*, 59–60, 68
    dating of 28
    plotting drift of India *80*, 80
    predicting true movement 23
magnetic time–scale **16**, 30, 105
magnetite, magnetized 11, 105
Main Boundary Fault *81*, 81, *82*
Main Central Thrust *81*, 81, *82*
mantle drag 108
marginal seas **56**
Marshall–Ellice Islands–Austral
seamount chain *21*, 22
mass deficit
    oceanic trenches *51*, 51–2
    relative, beneath central part of
        spreading ridges 35–6

median rift **34**, 34–5, 62
Mendocino Transform fault *14*, 45, 59, 66
metamorphic rocks **82**, 85, 107
metamorphism **54–5**, **82**
Meteor Rise 25
microcontinents **73**, 84, 107
Mid-Atlantic Ridge *34*, 34, 60
    magnetic anomaly profile *18*
    relative spreading rate 19–20, *20*
mid-ocean ridges **33**
    sites of sea-floor construction 13, 30
    symmetric magnetic anomaly patterns 15
    see also constructive plate boundaries
Moho 39, *53*, *76*, 84
    beneath the High Himalayas 81
    beneath oceanic crust 70
    see also petrological Moho; seismic Moho
mountain belts
    and active continental margins *see* Andes
    internal 79, 84
    linear 4
movement vectors, Pacific Plate **22**
Murray Transform 66

Nazca Plate 23, 43, *53*
    subduction of 77–8
Nazca–Cocos Plate system, expansion of 107
negative buoyancy force **88**
Newcastle coalfield, drift of 10, 97
Ninety-East Ridge 4, 16
    and hot-spot movement 25
North American Plate *53*, 53, 66
North Atlantic, time of formation 28
North Sea Basin *72*

ocean floor
    actively forming 33
    see also mid-ocean ridges
    depth of and failure planes 49
    entering Japan Trench 101
    height at constructive boundaries 18
ocean plate boundaries 33–61
    changing with time 63–8
    combined plate boundaries 33, **57**, 57–8
    conservative plate boundaries **17**, 18, **33**, **45**, 60, 61
    constructive plate boundaries **16**, 17, **33**, 33–42, *43*, 60, 88
    destructive plate boundaries **16–17**, 18, **33**, 46–57, 60, 88, 98
    transform faults and fracture zones 43–6
    triple junctions 58–60
ocean ridges 4, 88
ocean trenches 60, 61
    curved 46
    missing 54
    and plate margins 47
    siting of 46
ocean-floor bathymetry 18

oceanic crust 53
    age of 18, 28–9, 30, 65, 69, 97
    and preservation of 29, 99
    average density 69
    in back–arc basins 56
    basaltic 69
    and depth of ocean floor 49
    destruction of 16, 29, 30, 60, 75
    see also downgoing slabs
    formation of 13, 15, 36–9, 60, 101
    missing 29, 46, 99
    old, subducted 65
    origin of layered base 38–9
    structure of 36, 37, 99
    subducted and lost 107
oceanic drag force **87**, 90–1, 92
oceanic lithosphere
    deformation of 70
    flexure at subduction zone 50
oceanic plates **69**
    age of 28–9
    extending beneath landward plate 51–2
    rate of descent 95
    subduction of *see* subduction zones
ophiolites 36, 65, 107–8
Orozco fracture zone *41*
overlap basins **41**, *41*
overlaps 8
    caused by continental stretching 74–5
    caused by newer features 97
overriding plate resistance **89**

P-wave velocities 69, 84
Pacific margin 4
Pacific Ocean 28, 29, 46, 99
    predicting future shape 68, 107
Pacific Plate 43, 57, 91
    average rate of true motion 22
    boundaries of *65*, 65
    convergence with other plates 46
    minimum age for 52
    movement of 49, 100–1
    over hot spots 22–3
    movement vectors 22
    relative motion of 102
    subduction of 107
Pacific Plate–Antarctic Plate spreading ridge 19
Pacific Plate–North American Plate boundary 57
palaeoclimates, importance of 10–11
palaeolatitude indicators **12**, 29
    magnetite-bearing rocks as 12
palaeolatitudes 26, 105
palaeomagnetism 11–12, 29
    and apparent polar wander 12–13
    and sea-floor spreading 13–16
palaeopoles 12, 26
passive continental margins *72*, 72
passive margins 29
peridotite *39*, 39, 60
    layered *39*, 39
Peru–Chile Trench 46
petrological Moho **39**, 60
Philippine Plate 90

Pigafetta Basin 28
pillow lavas 36, 37, 41, 99
Pioneer Transform fault *14*, 45
plate boundaries 4, 30
    changes due to subduction 65–6, *67*
    case study: the Farallon Plate 66, *67*, 68, 106
    distinguishing between 62
    implications of sea-floor spreading 16–17
    oceanic 3
    within continental crust 3
    see also types of boundary
plate construction
    at spreading ridges 68
    see also constructive plate boundaries
plate convergence, a plate-driving force 78
plate margins, forces acting on 88–90
plate motion 7–31
    changes of at spreading ridges 63–4, 68
    controlled by a combination of forces 90
    modern, measurement of 27–8, 30
    relative 17–19, 30, 31, 90, 91
    significance of physical features 91–4, *92–3*
    true 19–27, 30, 90
plate movement vectors 49
plate movements 3
    comparison of 15–16
    true speed of 90–1
plate rotation, from palaeolatitudes 12, 97
plate tectonic cycles 70, 86, 107
plate tectonics, and continental crust 69–86
plate tectonics theory 3, 31
plate velocity 25, 108
    see also true plate velocity
plate–driving forces 78, 87–95
plates 3, 4, 31
    continental *see* continental plates
    dimensions and true velocities of *91*
    forces acting on bottom of 87–8
    moving apart 34
    not fixed in relation to asthenosphere 20
    oceanic *see* oceanic plates
    relative motion between 17–19, 30, 31, 90, 91
    rigidity of 50–1
plutons 55, *76*, 78, 107
    emplacement of *56*, 57, 76, 77
    Himalayan 80
polar wander path 26, *27*
    see also apparent polar wander; apparent polar wander path; true polar wander
Precambrian basement, Africa–S America 9–10, *10*
propagating rift model **64**
pyroclastic rock **55**

quadruple junction 66
quasars 27

radio telescopes 27
radiogenic heat 69
radiometric dating 16, 28
recycling
   of oceanic crust 69
   see also subduction zones
Red Sea 73, 74, 75
reference frames
   relative motion between 27
   see also hot-spot reference frame;
   magnetic reference frame
ridge crests
   active, igneous processes at 38–9, *39*
   over dyke development zones 41
ridge resistance **88**
ridge segments 39–42, 43–4, 60
ridge (tip) propagation *64*, 64, 68, 74,
102
   by magma pulses 42, *43*
ridge–push force 87, **88**, 90, 95, 103
   as a driving force 94
ridge–transform intersections 46, 100
rift valleys **71**
rock deformation **54**
RRR triple junctions 58
   existence as 'fossils' 59, 61
RTI *see* ridge–transform intersections

St Helena 25
St Vincent *53*
San Andreas Fault zone 47, 59, 66, 68,
73, 83, 85
   movement along 28
satellite laser ranging 27–8
Scottish Highlands 86, 108
sea-floor spreading **13**
   asymmetric **63**
   implications for plate boundaries 16–
   17
   see also mid-ocean ridges; spreading
   axes; spreading ridges
seamount chains, prediction of 23–5
seamounts 4, *21*, 21, 23, 25
sediment accumulation
   at subduction zones 54, *71*, 72
   in graben 84
   see also accretionary prisms
sediment cover, deep ocean 54
sedimentary rocks, Andean 77
seismic Moho **39**, 60, 69
shear zones *72*, *76*, 84
sheeted dyke complexes 36–7, 70, 99
Shona hot spot trail 25
Siqueiros fracture zone *41*
slab resistance **88**, 90
slab–pull force **88**, 90, 94, 95, 103, 108
   an effective driving force 94
Smithsonian Map 3, 4, 19, 22, 23, 33, 43,
46, 47, 82, 98
South America 77
   palaeolatitude variations 12, *13*
   see also Africa–S. America,
   geological fit
South American Plate 108

South Atlantic, average rate of relative
spreading 18
South Indian Ridge 90
Southern Ocean continents *79*, 79
   best fit *8*, 8, *9*
spreading axes 33, 34, 60, 72, 101
   active, subduction of 65–6
   Gulf of Aden *74*, 74
   and the Red Sea 83
   implications of *17*
   offset 43
   traced onshore 82–3, 85
spreading centres **33**, 60
   overlapping **40–1**, *41*, 41–2
   and ridge segments 39–42, *43*
spreading direction, change in 63–4, 68
spreading rate 16, **17**, 18, 30, 60, 98
   controls central portion of spreading
   ridges 34–5
   relative 19, *20*
spreading ridges **33**, 70
   adjustment by rift propagation *64*, 64,
   68
   adjustment by rotation *63*, 63–4, 68
   changes at with time 63–4
   changes of plate motion at 63–4
   containing lower density material 35
   movement of 25
   offsets on 40, 43
   processes at 36–9
   and ridge topography 34–9
   segmentation of 40
   siting of spreading centres 40
   see also transform faults
stable junctions **58**, 106
stress
   compressional 51, 56, 57
   tensional (tensile) 51, 56, 57
stress relaxation 84
strike–slip fault systems 85
subducted slab *see* downgoing slab
subduction
   angle of 48–9, 55, 90
   case study: the Andes 77–9
   not always perpendicular to trenches
   57–8
subduction complex, position of island
arc volcanoes 55
subduction zones **46**, 47–52, 60, 61, 108
   case study: Lesser Antilles 52–7
subduction–transform complex 106
subsidence, surface *72*, 72

Taurus Mountains 83
tension 46, 51, 57
thermal relaxation *71*, *72*, 84
thrusts/thrust faults **54**, *76*, **77**, *78*, 84,
107
Tobago Trough *53*
Tonga Trench 55, 101
transform fault resistance **88**, 103, 108
transform faults **43**, 43–5, *53*, 60, 62, 63,
88
   at a destructive margin 57–8, *58*
   case study: Clipperton transform fault
   45–6

counterpart within continental crust
82
differing from transcurrent faults 45
   geometry of 44, 100
   Gulf of Aden *74*
   as topographic features 45
   see also continental transforms; leaky
   transforms
transform plate boundary 102
transform zone 46
trench suction force 57, 87, **89**, 103
trench–arc complexes **46**, *51*, 51–2, 55
   see also island arcs; ocean trenches
trench–slab system migration 55–6
triple junctions **58**, 58–60, 61, 62, 63,
106
   evolution of *58*, 102
   and Farallon Plate 66
   types of 59, 106
Tristan da Cunha hot-spot trail 25, 105
true plate motion 19–27, **20**, 30, 90
   of adjacent plates 23
   calculated in terms of latitude 26
true plate vectors, using apparent polar
wander path 26
true plate velocity 91, 94
   average, correlation with other
   variables 93–4
   depending largely on slab–pull 94
true polar wander **26**, 30
   and magnetic reference frame 25–7
   see also hot-spot reference frame
TTT triple junctions 58–9
turbidites 53, 101

underthrusting 78
unstable junctions **58**, 58–9, 106

very long baseline interferometry (VLBI)
27, 105
Vine–Matthews hypothesis 15–16
volcanic island chains *21*, 21, 22, 25, 105
   island arcs **46**, **47**, 61
   possible mode of formation *21*
volcanic vents, active, spreading ridges
34
volcanism 61
   island arc 55, 107
   on landward side of trenches 46
   predicted age of, hot–spot tracks 25
   and within–continent divergence 72
volcanoes 4, 34, 77
   active 47, 62
   hot–spot 98
   island arc 55

Wadati–Benioff zones **49**, 60, 62, 105
Walvis Ridge 25
Wilson, J.T. 44–5

Zagros fold belt 83
Zangbo suture *82*

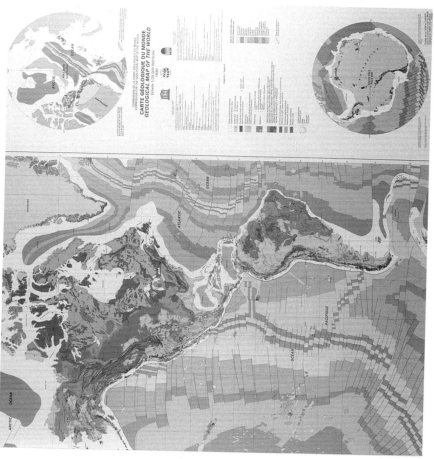

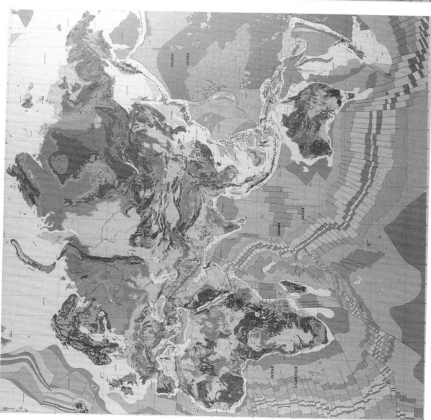

Plate 2.1 Age of the ocean floor, showing strips of different ages derived mainly from measurements of magnetic anomaly stripes. Ages range from reds (0–20 Ma) through yellows (20–35 Ma), oranges (35–80 Ma), greens (80–140 Ma) and light blue (140–160 Ma) to the oldest oceanic crust at between 160 and 170 Ma. Oceanic crust this old only occurs on the eastern seaboard of the USA and east of the Philippines in the Pacific Ocean. (*UNESCO*)

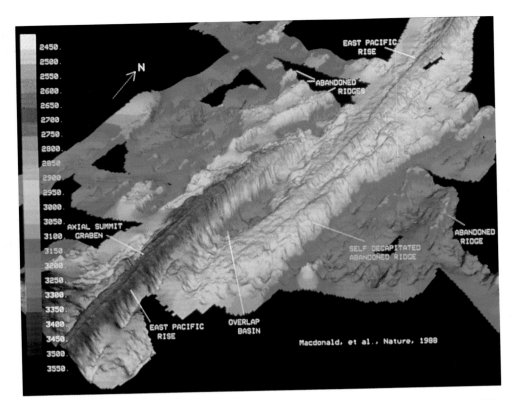

Plate 2.2 *Sea Beam* sonar image of the East Pacific Rise at 11° 50′ N, looking northwest. The area shown is approximately 80 km by 65 km. An axial summit graben is prominent in the foreground on the southern ridge segment and in the background on the northern ridge segment. A ridge that will shortly be abandoned lies on the eastern flank of the northern segment. Its strike relative to the active ridge suggests that it has been propagating southwards at about 250 mm yr⁻¹. (*Prof. K. MacDonald, University of California, Santa Barbara*)

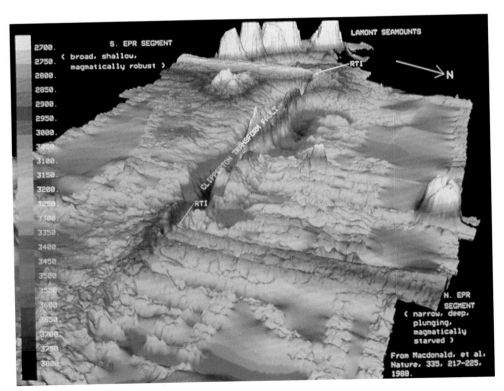

Plate 2.3 *Sea Beam* sonar image of the East Pacific Rise at about 9° 45′ N, looking west-southwest. The area shown is approximately 100 km by 175 km. The view shows the Clipperton Transform Fault and adjacent segments of the East Pacific Rise. The transform fault itself is a narrow cleft cutting through rugged relief in the valley joining two ridge–transform intersections. The East Pacific Rise is displaced by about 85 km. Sonar data coverage is nearly complete, but areas of very smooth contours are computer interpolations. (*Prof. K. MacDonald, University of California, Santa Barbara*)

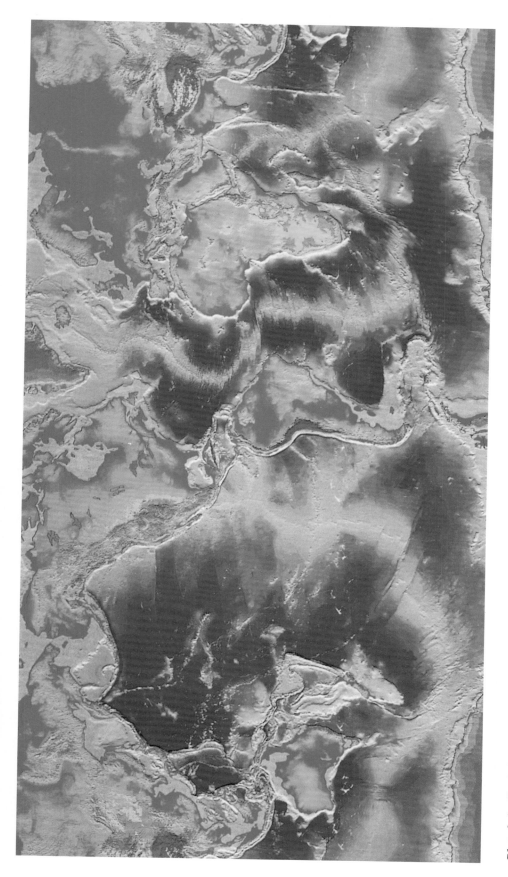

Plate 2.4 Shaded relief map of the Earth's solid surface. (*NASA*)

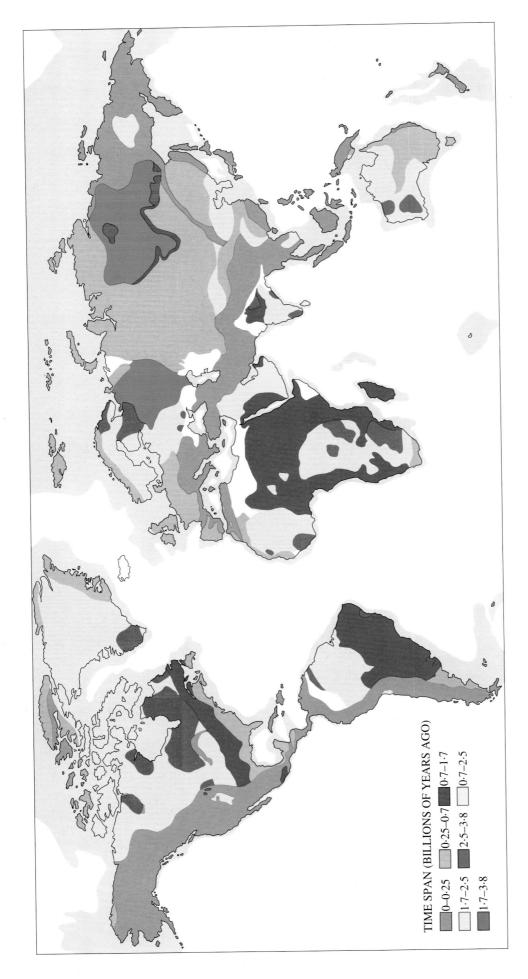

TIME SPAN (BILLIONS OF YEARS AGO)

0–0.25   0.25–0.7   0.7–1.7

1.7–2.5   2.5–3.8   0.7–2.5

1.7–3.8

Plate 2.5   Continental crust is a mosaic of belts of rock, deformed at very different times during the Earth's history. Note the long, linear belts that make up the recent mountain chains (0–0.25 billion years), whereas older mountain chains have been broken up by plate movements. There is some overlap of ages in the key because different belts have different geological histories. (Ignore the areas of land shown in white.) (*By A. Tomko in B. Clark (1983) 'The continental crust', Scientific American, September © 1983 W. H. Freeman Inc.*)